Algebra with Galo

Courant Lecture Notes in Mathematics

Emil Artin
Notes by Albert A. Blank

15

Algebra with Galois Theory

Courant Institute of Mathematical Sciences
New York University
New York, New York

American Mathematical Society
Providence, Rhode Island

2000 *Mathematics Subject Classification.* Primary 12–01, 12F10.

Library of Congress Cataloging-in-Publication Data

Artin, Emil, 1898–1962.
Algebra with Galois theory / E. Artin, notes by Albert A. Blank.
p. cm. — (Courant lecture notes ; 15)
ISBN 978-0-8218-4129-7 (alk. paper)
1. Galois theory. 2. Algebra. I. Blank, Albert A. II. Title.

QA214.A76 2007
512—dc22 2007060799

Printed in the United States of America.

∞ The paper used in this book is acid-free and falls within the guidelines established to ensure permanence and durability.
Visit the AMS home page at `http://www.ams.org/`

10 9 8 7 6 5 4 3 2 1 12 11 10 09 08 07

Contents

Editors' Note

Because what was in 1947 "modern" has now become standard, and what was then "higher" has now become foundational, we have retitled this volume *Algebra with Galois Theory* from the original *Modern Higher Algebra. Galois Theory.*

Jalal Shatah, *Executive Editor*
Paul Monsour, *Managing Editor*
August 2007

CHAPTER 1

Groups

We concern ourselves with sets G of objects $a, b, c, \ldots$ called elements. The sentence "a is an element of G" will be denoted symbolically by $a \in G$. Assume an operation called "multiplication" which assigns to an ordered pair of objects a, b of G another object $a \cdot b$ (or simply ab) the *product* of a and b. It is useful to require that G be *closed* with respect to multiplication, namely:

(1) If $a, b \in G$, then $a \cdot b \in G$.

EXAMPLES.

(a) Let G be the set of positive integers. If subtraction is taken as the "multiplication" in G, then G is certainly not closed, e.g., $3 \cdot 5 = 3 - 5 = -2$. If taking the greatest common divisor is our multiplication, then closure is obvious.

(b) Take G to be the set of functions of one variable. If $f(x), g(x) \in G$ define $f(x) \cdot g(x) = f[g(x)]$, e.g., $e^x \cdot \log x = e^{\log x} = x$.

EXERCISE 1. Write out the multiplication table and thereby show closure for the set of functions

$$f_1 = x, \quad f_2 = \frac{1}{x}, \quad f_3 = 1 - x, \quad f_4 = \frac{1}{1-x}, \quad f_5 = \frac{x}{x-1}, \quad f_6 = \frac{x-1}{x}.$$

SOLUTION.

	f_1	f_2	f_3	f_4	f_5	f_6
f_1	f_1	f_2	f_3	f_4	f_5	f_6
f_2	f_2	f_1	f_4	f_3	f_6	f_5
f_3	f_3	f_6	f_1	f_5	f_4	f_2
f_4	f_4	f_5	f_2	f_6	f_3	f_1
f_5	f_5	f_4	f_6	f_2	f_1	f_3
f_6	f_6	f_3	f_5	f_1	f_2	f_4

where $f_i \cdot f_j$ is listed in the ith row and j^{th} column.

We make the further requirement that multiplication obey the *associative law*:

(2) If $a, b, c \in G$, then $(ab)c = a(bc)$. This is a rather strong condition. It is not generally satisfied; consider, e.g., subtraction among the integers. For functions of one variable, as above, it is valid, however. If $f(x)$, $g(x)$, $h(x)$ are any three functions we have

$$(fg)h = f(g(h(x))) = f(gh).$$

EXERCISE 2. Deduce the associative law for four elements from (2), that is, show that the five possible products of four elements written in a given sequence are all equal. Furthermore, attempt to determine the number of possible products of n elements given in a linear order. For example, the elements a_1, a_2, a_3, a_4 in that the order yield the products $(a_1a_2)(a_3a_4)$, $a_1(a_2(a_3a_4))$, etc. *Hint*: Let α_n be the number of products of $a_1, a_2, \ldots, a_n$. Find a recursion formula for α_n and use the Lagrange generating function

$$f(x) = \alpha_1 x + \alpha_2 x^2 + \cdots + \alpha_n x^n + \cdots .$$

EXERCISE 3. The associative law for n elements states that all possible products of n elements written in a prescribed order, e.g., $a_1, a_2, \ldots, a_n$, yield the same result. Prove the associative law for any number of elements using only (2) (the associative law for three elements).

PROOF FOR EXERCISE 3: We assume the validity of the associative law for all products of m factors, $m \leq n$, and show that this implies the validity of the law for $n+1$. Consider the particular product $(n+1)\prod_{k=1}^{n+1} a_k$ which is obtained from the $n+1$ elements $a_1, a_2, \ldots, a_{n+1}$ by successively multiplying on the right, i.e.,

$$\prod_{k=1}^{1} a_k = a_1,$$

$$\prod_{k=1}^{n+1} a_k = \left(\prod_{k=1}^{n} a_k\right) a_{n+1}.$$

Let P_{n+1} be any product of the $n+1$ elements $a_1, a_2, \ldots, a_{n+1}$ taken in that order. Since P_{n+1} is the result of at least one multiplication, we may write

$$P_{n+1} = P_1^m P_{m+1}^{n+1}, \quad 1 \leq m \leq n,$$

where P_1^m is some product of the elements $a_1, a_2, \ldots, a_m$ in that order and P_{m+1}^{n+1} of the remaining elements $a_{m+1}, a_{m+2}, \ldots, a_{n+1}$. By the induction hypothesis we have

$$P_\mu^\nu = \prod_{k=\mu}^{\nu} a_k$$

for any μ, ν such that $\nu - \mu + 1 \leq n$. Specifically, we have

$$\begin{aligned} P_{n+1} = \left(\prod_{j=1}^{m} a_j\right)\left(\prod_{k=m+1}^{n+1} a_k\right) &= \prod_{j=1}^{m} a_j \cdot \left[\left(\prod_{k=m+1}^{n} a_k\right) \cdot a_{n+1}\right] \\ &= \left(\prod_{j=1}^{m} a_j \prod_{k=m+1}^{n} a_k\right) \cdot a_{n+1} \\ &= \left(\prod_{k=1}^{n} a_k\right) \cdot a_{n+1} = \prod_{k=1}^{n+1} a_k, \end{aligned}$$

each step being a simple application of (2). □

1.1. The Concept of a Group

A set G will be called a *group* if it satisfies the following conditions:

(1) *Closure.* There exists an operation called multiplication which assigns to any ordered pair $a, b \in G$ a product $ab \in G$.
(2) *Associative Law.* If $a, b, c \in G$ then $(ab)c = a(bc)$.
(3) *Identity.* There exists an $e \in G$, called the (left) *identity*, such that $ea = a$ for all $a \in G$.
(4) *Inverse.* For every $a \in G$ there is an $a^{-1} \in G$, called the (left) *inverse* of a, such that $a^{-1}a = e$.

Let us examine the product

$$(a^{-1})^{-1}a^{-1}aa^{-1}.$$

On one hand,

$$[(a^{-1})^{-1}a^{-1}][aa^{-1}] = e[aa^{-1}] = aa^{-1},$$

and on the other

$$[(a^{-1})^{-1}][(a^{-1}a)a^{-1}] = [(a^{-1})^{-1}][ea^{-1}] = (a^{-1})^{-1}a^{-1} = e.$$

Consequently,

$$aa^{-1} = e.$$

The existence of the left inverse implies the existence of a right inverse. A similar result holds for the identity; for consider the product

$$aa^{-1}a.$$

First we have

$$aa^{-1}a = (aa^{-1})a = ea = a.$$

But also

$$aa^{-1}a = a(a^{-1}a) = ae.$$

Consequently,

$$ae = a,$$

and the existence of the right identity implies the existence of a left identity.

EXERCISE 4. Two systems of postulates are said to be equivalent if either system can be derived logically from the other. Show that the system (1), (2), (3), (4) is equivalent to the system in which (3) and (4) are replaced by:

(3′) There is a right identity $e \in G$ such that $ae = a$ for all $a \in G$.
(4′) To each $a \in G$ there is a right inverse $a^{-1} \in G$ such that $aa^{-1} = e$.

Apparently the words *right* and *left* need not be included in (3), (4), (3′), or (4′).

EXERCISE 5. Consider the postulate system in which (3) and (4) are replaced by:

(3*) There exists a left identity $e \in G$; that is, $ea = a$ for all $a \in G$.
(4*) To each $a \in G$ there is a right inverse $a^{-1} \in G$; that is, $aa^{-1} = e$. Determine whether this system of postulates defines a group. If not, give a counterexample.

SOLUTION. For any $a \in G$ define multiplication by $ax = x$ for all $x \in G$. This system satisfies the postulates (1), (2), (3*), and (4*). What group property does it not satisfy?

For ordinary numbers, the quotient $b \div a$ of two numbers can be defined as the solution of the equation $ax = b$. Consider similar equations for elements of G:

$$\text{(a)}\ ax = b, \quad \text{(b)}\ xa = b, \quad \text{(c)}\ axb = c$$

If (a) is true for some x, then

$$a^{-1}ax = a^{-1}b = ex = x.$$

Hence, if there is a solution, it is $a^{-1}b$ and it is therefore unique; $a^{-1}b$ is in fact a solution. Similar reasoning shows that (b) possesses the unique solution $x = ba^{-1}$ and (c) the unique solution $a^{-1}cb^{-1}$. The existence of a unique solution for each of the above equations demonstrates a property of the group analogous to division.

Since a^{-1} is the solution of the equation $xa = e$, a^{-1} is unique. Similarly, e is the unique solution of $xa = a$. We observe that the solution of $x(ab) = e$ is $(ab)^{-1} = b^{-1}a^{-1}$. In general, the inverse of a product

$$(a_1a_2 \cdots a_n)^{-1} = a_n^{-1} \cdots a_2^{-1}a_1^{-1}.$$

If $x = (a^{-1})^{-1}$, then x satisfies the equation $xa^{-1} = e$, which has the unique solution $x = a$. Thus the inverse of the inverse of an element is the element itself.

EXERCISE 6. Show that postulates (3) and (4) may be replaced by

(3^+) If $a, b \in G$, the equations

$$xa = b, \quad ay = b,$$

possess (not necessarily unique) solutions $x, y \in G$.

A group that satisfies the *commutative* law,

(5) If $a, b \in G$, then $ab = ba$,

is said to be commutative or abelian.

EXERCISE 7. Show that the six functions of Exercise 1 form a noncommutative group with respect to their rule of multiplication. Determine the identity element and the inverse to each function.

1.2. Subgroups

If G is a group and S is a subset of G that is itself a group under the same operation as G, then S is called a *subgroup* of G.

EXAMPLE. Take G to be the set of rational numbers other than zero under ordinary multiplication. G has, e.g., the subgroups

(a) the positive rational numbers
(b) the powers of any element
(c) the set consisting of $+1$ and -1

Trivially, (d) the set G itself or (e) the set consisting of the element 1.

THEOREM 1.1 *Necessary and sufficient conditions for a subset S of G to be a subgroup are*:

(i) Closure. *If $s_1, s_2 \in S$, then $s_1 s_2 \in S$.*
(ii) Inverse. *If $s \in S$, then $s^{-1} \in S$.*

PROOF: *Necessity.* If S is a subgroup, (i) holds by definition. The identity $e \in S$ by the uniqueness in G and existence in S by the solution of the equation

$$xs = s.$$

Note that (ii) is similarly established through the equation $xs = e$.

Sufficiency. If (i) and (ii) hold, then S is a subgroup. From (ii) if $s \in S$ then s^{-1} is an element of S and hence (i) gives $e \in S$. The associative law holds for elements of S since they are elements of G. The proof of the theorem is complete. □

If S is a subgroup of G and $a \in G$, the *coset* aS is defined to be the set of all elements $a \cdot s$, where $s \in S$.

EXAMPLE. Take for G the set of all rational numbers excluding zero under ordinary multiplication. Let S be the set of all positive elements of G. There are only two cosets, S and $-S = -1S$. These have no elements in common and both sets together cover G. If we take instead $S = \{+1, -1\}$ then the cosets are $aS = \{+a, -a\}$. Here the same coset is given by $+a$ and $-a$. Note again that no two cosets overlap and that the cosets cover G. These results are valid in general.

Let S be a subgroup of G and take $a, b \in G$.

LEMMA 1.2 *If the cosets aS and bS have an element c in common, then $aS = bS$.*

Assume for some $s, s' \in S$ we have $c = as = bs'$. Therefore $b = as(s')^{-1}$. From Theorem 1.1, $s(s')^{-1} = s'' \in S$ and consequently $bS = as''S$. Now $s''S = S$, since if we suppose S to be any group, s any element of S, we have

$$sS \subset S$$

(read: sS is a subset of S, or all elements of sS are elements of S). Also,

$$s^{-1}s \subset S \quad \text{or} \quad S \subset sS.$$

Therefore

$$sS = S.$$

In the above argument we may now write $bS = as''S = aS$.

LEMMA 1.3 *Every $a \in G$ is contained in some coset*

$$a \in aS \quad \textit{since} \quad e \in S$$

and hence $ae = a \in aS$. G is covered by the cosets of S.

If G is a finite group, then the number of its elements is called the *order* of G.

THEOREM 1.4 *Let G be a finite group of order N and S a subgroup of order n. The number n of elements in the subgroup is a divisor of N.*

PROOF: The cosets aS have the same number of elements as S. For let S consist of the distinct elements $s_1, s_2, \ldots, s_n$. aS consist of $as_1, as_2, \ldots, as_n$, where

$$as_1 \neq as_k, \quad i \neq k.$$

For otherwise we would have $as_i = as_k$ and hence $s_i = s_k$, $i \neq k$, contrary to the definition of the s_i.

Consequently, aS consists of exactly n elements. Let j be the number of cosets. By Lemmas 1.2 and 1.3 the cosets cover G without overlapping. It follows that

$$N = jn. \qquad \square$$

Take $a \in G$. We denote aa by a^2 or, in general, we define all the integral powers a^μ of a by

$$\begin{aligned} a^\mu &= aa \cdots a \ (\mu \text{ times}) \text{ for } \mu > 0, \\ a^0 &= e, \\ a^\mu &= a^{-1}a^{-1} \cdots a^{-1} \ (-\mu \text{ times}) \text{ for } \mu < 0. \end{aligned}$$

The set of all powers of a is a group and clearly the smallest group containing a. The problem of determining the smallest group containing as few as two elements is already of an entirely different nature. For example, what can be said about

$$(ab)^n = ab \cdot ab \cdots ab \ (n \text{ times})?$$

If multiplication is commutative such products can be handled, but this does not apply in general.

EXERCISE 8. Show that the powers of elements obey the usual properties of exponents

$$\begin{aligned} a^\mu a^\nu &= a^{\mu+\nu}, \\ (a^\nu)^\mu &= a^{\nu\mu}. \end{aligned}$$

The first property implies the commutative law for multiplication of powers of a.

The set S of all powers of a forms a subgroup since S is closed under multiplication and inverses exist (cf. Theorem 1.1).

Case 1. The powers of a are all distinct. S is then called an *infinite cyclic group*.

Case 2. There exist integers i and k with, say, $i < k$ such that $a^i = a^k$. Multiplying on both sides by a^{-i} we obtain $e = a^{k-i}$. Thus the set of positive integers for which $a^\mu = e$ is not empty. Let d be the smallest such integer

$$a^d = e \Rightarrow a^{qd} = e \quad \text{for all integers } q$$

(read: "implies" for "$\Rightarrow$"). Conversely, if $a^m = e$, m is a multiple of d, for we may write $m = qd + r$ where $0 \leq r < d$

$$a^r = a^{m-qd} = a^m a^{-qd} = e.$$

But d is the smallest positive integer for which $a^d = e$. Hence r must be zero whence $m = qd$. The powers

$$a^0, a^1, a^2, \ldots, a^{d-1}$$

are all distinct for otherwise we would have

$$a^i = a^k, \quad 0 \le i < k < d, \qquad \text{or} \qquad a^{k-i} = e;$$

this equation is impossible for $0 < k - 1 < d$. Any other power of a must be equal to one of these, for example $a^d = e$, $a^{d+1} = a, \ldots$, or, in general,

$$a^{qd+r} = a^r, \quad 0 \le r < d.$$

Thus there are only d distinct powers of a. S is called a *cyclic* subgroup of order d and d is called the *period* of a.

THEOREM 1.5 *The period of any element of a finite group is a divisor of the order of the group.*

PROOF: This is an immediate consequence of Theorem 1.4. Let G be a finite group of order N and a any element of G. If d is the period of a, we may write $N = dj$. From $a^d = e$ we have

$$a^{dj} = a^N = e.$$

This statement for prime N is equivalent to Fermat's theorem in arithmetic. □

COROLLARY *If the order of G is p, a prime, then G must be cyclic.*

PROOF: The period of any element must be a divisor of p and is therefore either p or 1. The only element of period 1 is e. Consequently, if $a \in G$ and $a \neq e$ the period of a must be p. □

There is "essentially" one cyclic group of order n. Phrased differently, two cyclic groups of the same order have the "same structure." The notion of "same structure" will be examined later in more detail.

EXAMPLES. Let us determine all possible structures of groups of order 4. The period of any element must be 1, 2, or 4. If there is an element a of period 4 then e, a, a^2, a^3 exhaust the group. On the other hand, if there is no element of period 4, then all elements but e must have the period 2. Thus if e, a, b, c denote the different elements of the group we have $a^2 = b^2 = c^2 = e$. Consider the element $x = ab$. From $ax = aab = b$ we have clearly $x \neq e$, $x \neq a$. From the uniqueness of the solution $y = e$ of the equation $yb = b$ it follows that $x \neq b$. Therefore x must be c. The commutative law holds in this group, for if $x \in G$ then $x = x^{-1}$ and consequently $ab = (ab)^{-1} = b^{-1}a^{-1} = ba$. It is a simple matter to write out the multiplication table:

	e	a	b	c
e	e	a	b	c
a	a	e	c	b
b	b	c	e	a
c	c	b	a	e

We have shown that there are essentially two groups of order 4 and both are commutative.

Groups of order 6 are essentially of two kinds, the cyclic group and the noncommutative group given in Exercise 1. This last is the simplest example of a noncommutative group. One of the unsolved problems of algebra is that of classifying all the groups of order n. There is, of course, always the cyclic group of order n and for n prime, only the cyclic group. For nonprimes there is no general theory although a classification has been achieved for special cases. The table below gives a summary for the first few cases:

N	4	6	8	9	10	12	14	15
μ	2	2	5	2	2	5	2	1
ν	0	1	2	0	1	3	1	0

where μ is the total number and ν the number of noncommutative groups of order N.

EXERCISE 9. The two noncommutative groups of order 8 are essentially:

(a) The symmetries of the square, i.e., the rotations in space which take the square into itself.

(b) The group formed by the quaternion units $\pm 1, \pm i, \pm j, \pm k$.

Construct the multiplication table for those two groups and show that they do not have the same structure.

(a) The symmetries of the square.

If a rotation replaces the vertices (1234) by the vertices $(a_1a_2a_3a_4)$, then denote the rotation simply by $(a_1a_2a_3a_4)$. The identity is clearly $e = (1234)$. Denote by $a = (2341)$ the counterclockwise rotation through 90°. Let $a^2 = b = (3412)$ and $c = a^2 = (4123)$. We have $a^4 = e$. The powers of a form a group S of order 4. If s denotes a rotation of 180° about the axis 1–3 we have $s = (1432)$. The coset sS is simply

$$s = (1432), \qquad sa = (2143) = t,$$

$$sa^2 = (3214) = u, \quad sa^3 = (4321) = v;$$

these together with the powers of a exhaust the symmetries of the square:

	e	a	b	c	s	t	u	v
e	e	a	b	c	s	t	u	v
a	a	b	c	e	v	s	t	u
b	b	c	e	a	u	v	s	t
c	c	e	a	b	t	u	v	s
s	s	t	u	v	e	a	b	c
t	t	u	v	s	c	e	a	b
u	u	v	s	t	b	c	e	a
v	v	s	t	u	a	b	c	e

(b) The quaternion group.

This is obtained at once by the ordinary rules of multiplication of the quaternion units

	$+1$	$+i$	$+j$	$+k$	-1	$-i$	$-j$	$-k$
$+1$	$+1$	$+i$	$+j$	$+k$	-1	$-i$	$-j$	$-k$
$+i$	$+i$	-1	$+k$	$-j$	$-i$	$+1$	$-k$	$+j$
$+j$	$+j$	$-k$	-1	$+i$	$-j$	$+k$	$+1$	$-i$
$+k$	$+k$	$+j$	$-i$	-1	$-k$	$-j$	$+i$	$+1$
-1	-1	$-i$	$-j$	$-k$	$+1$	$+i$	$+j$	$+k$
$-i$	$-i$	$+1$	$-k$	$+j$	$+i$	-1	$+k$	$-j$
$-j$	$-j$	$+k$	$+1$	$-i$	$+j$	$-k$	-1	$+i$
$-k$	$-k$	$-j$	$+i$	$+1$	$+k$	$+j$	$-i$	-1

The two groups do not have the same structure since the group of symmetries has 5 elements of period 2 while the quaternion group has only one such element.

CHAPTER 2

Rings and Fields

In the chapter on groups we have isolated certain properties of ordinary multiplication of numbers and examined these in some detail. It has become obvious that the notion of group and multiplication in a group is a far more general concept and has many more applications than that of multiplication of numbers. It is now our purpose to define systems which will include some of the ordinary properties of numbers (e.g., addition, multiplication, and later, division). At the same time these systems will remain sufficiently general to have wider application. Consider first a set for whose elements two operations called "addition" and "multiplication" are defined.

EXAMPLE. The addition and multiplication of odd and even integers obeys the rules Even + Odd = Odd, Even × Even = Even, etc. The total behavior of addition and multiplication of Even and Odd is given in the tables:

+	Even	Odd
Even	Even	Odd
Odd	Odd	Even

×	Even	Odd
Even	Even	Even
Odd	Even	Odd

If "Even" is replaced by the number 0 and "Odd" by the number 1, these tables are the same as for ordinary addition and multiplication, together with the special rule $1 + 1 = 0$.

Consider a set T which is closed with respect to two operations, addition and multiplication. The element resulting from the addition of two elements is called the *sum* $a + b$. We postulate that

(I) The elements of T are a group under addition.

The identity of this additive group is denoted by 0. The inverse of the element a is denoted by $-a$. According to the customary convention, the element $a + (-b)$ is written $a - b$. The rules for the use of the minus sign before parentheses are easily demonstrated:

$$\begin{aligned} a - (b + c - d) &= a + \{-[b + c + (-d)]\} \\ &= a + [-(-d) + (-c) + (-b)] = a + d - c - b. \end{aligned}$$

Note that the order of the elements in the parenthesis has been reversed. If the commutative law holds the elements may be written in any order.

(II) The distributive laws.

If $a, b, c \in T$, then

[a] $$a(b+c) = ab + ac$$

[b] $$(b+c)a = ba + ca$$

Consider the product $(a+b)(c+d)$. From II[a] and then II[b] we have

$$(a+b)(c+d) = (a+b)c + (a+b)d = ac + bc + ad + bd$$

and from II[b] and then II[a]

$$(a+b)(c+d) = a(c+d) + b(c+d) = ac + ad + bc + bd.$$

Setting the two results equal yields

$$bc + ad = ad + bc.$$

The distributive laws imply that elements which are products are commutative with respect to addition. Thus only if some elements are not products—and this case seldom occurs—is addition noncommutative. Consequently, instead of (I) we take

(I*) The elements of T form a commutative group under addition.

Now let us consider some other consequences of (II). We have

$$ab = a(b+c) = ab + ac$$

whence $a0 = 0$ for all a.

In a similar way we can show

$$0a = 0.$$

The product of any element with zero is zero.

It is also possible to prove the usual rules concerning the minus sign in multiplication

$$a[b + (-b)] = \begin{cases} a0 = 0 \\ ab + a(-b) \end{cases}$$

or $ab + a(-b) = 0$. Therefore $a(-b) = -ab$. By a similar proof

$$(-a)b = -ab.$$

From the combination of these results we have

$$(-a)(-b) = -((-a)b) = -(-(ab)) = ab.$$

In some of the literature a set T satisfying (I*) and (II) is called a *ring* and mention is made of "associative rings," i.e., rings which satisfy the postulate

(III) $a, b, c \in T \Rightarrow a(bc) = (ab)c$.

We adopt a more customary usage and define a *ring* to be a set which is an "associative ring" in the sense above. A ring, then, is a set, closed with respect to addition and multiplication, that is a commutative group with respect to addition and obeys the distributive law of multiplication over addition and the associative law of multiplication.

EXAMPLE. The integers under ordinary addition together with the rule that the product of any two elements is zero. Any commutative group can be made to furnish a ring in this manner.

We are also interested in rings which possess division properties, i.e., inverses with respect to multiplication. Hence we introduce the postulate

(IV) The set T, excluding zero, is a group under multiplication.

The multiplicative identity is hereafter denoted by 1. A set F which satisfies (I), (II), and (IV) is called a *field*. Thus a field is a group with respect to addition, satisfies the distributive law of multiplication over addition, and, except for the additive identity, is a group with respect to multiplication. Clearly (IV) implies (III). Furthermore (I*) holds in this system since the existence of the multiplicative identity means that every element can be expressed as a product. Consequently, addition is commutative as a consequence of (II). From (IV) we also obtain the "cancellation law"

$$a \neq 0,\ b \neq 0 \Rightarrow ab \neq 0$$

since zero is not an element of the multiplicative group.

A *commutative field* is a field which obeys the commutative law of multiplication. If a field is commutative, it is convenient to adopt the notation for fractions. We write a/b for $b^{-1}a$ where $b \neq 0$. Thus $a/b = c/d \Leftrightarrow b^{-1}a = d^{-1}c \Leftrightarrow ad = bc$. From this rule immediately follows

$$\frac{ca}{cb} = \frac{a}{b}.$$

EXERCISE 1. Derive the usual rules for addition, multiplication, and taking the reciprocals of fractions.

2.1. Linear Equations in a Field

Consider the system of m linear equations in n unknowns

$$[1] \qquad \begin{cases} L_1 = a_{11}x_1 + a_{12}x_2 + \cdots + a_{1n}x_n = b_1 \\ L_2 = a_{21}x_1 + a_{22}x_2 + \cdots + a_{2n}x_n = b_2 \\ \vdots \\ L_m = a_{m1}x_1 + a_{m2}x_2 + \cdots + a_{mn}x_n = b_m \end{cases}$$

where the coefficients a_{ik} and b_j are elements of a field F. The system [1] is said to have a *solution* in F if there exist $c_1, c_2, \ldots, c_n \in F$ such that [1] is a true statement when c_i is substituted for x_i. If the b_j are all zero [1] is said to be a system of homogeneous equations. A system of homogeneous equations clearly has the *trivial* solution all $x_i = 0$. Any other solution is called nontrivial.

THEOREM 2.1 *If $n > m$, the system*

$$[2] \qquad L_i = a_{i1}x_1 + a_{i2}x_2 + \cdots + a_{in}x_n = 0 \quad (i = 1, 2, \ldots, m)$$

of m homogeneous equations in n unknowns always has a nontrivial solution.

REMARK. The condition that the equations be homogeneous is quite necessary since, for example, the equations

$$x + y + z = 1, \quad x + y + z = 0,$$

can have no solution in F.

PROOF: We use induction on m.

(1) If $m = 0$ the theorem certainly holds since we have $n > 0$ unknowns and no conditions on them. We could take all $x_i = 1$.
(2) Assume the theorem is true for all systems in which the number of equations is less then m.

Case 1. All $a_{ik} = 0$. The theorem is true, for we may choose all $x_i = 1$.

Case 2. There is a nonzero coefficient. Without loss of generality we may assume specifically $a_{11} \neq 0$ since altering the order of the x_i or of the equations has no effect upon the existence or nonexistence of solutions. We take $a_{11} = 1$ since we may multiply on the left by a_{11}^{-1}. Let us examine the system of equations

$$[3] \qquad \begin{cases} L_1 = 0 \\ L_2 - a_{21}L_1 = 0 \\ \quad\vdots \\ L_m - a_{m1}L_1 = 0 \end{cases}$$

obtained by "eliminating" the variable x_1 from the last $m - 1$ equations in [2]. Any solution of [2] is obviously a solution of [3]. Conversely any solution of [3] is a solution of [2] since the solution must satisfy $L_1 = 0$. It suffices to show that [3] has a nontrivial solution.

The system of equations

$$[3'] \qquad \begin{cases} L_2 - a_{21}L_1 = 0 \\ \quad\vdots \\ L_m - a_{m1}L_1 = 0 \end{cases}$$

is essentially a system of $m - 1$ equations in the $n - 1$ unknowns $x_2, x_3, \ldots, x_m$. From the induction assumption this system possesses a nontrivial solution. Using this solution we complete the solution of [3] by substituting the first equation to obtain x_1. The proof of the theorem is in no way changed when the coefficients are multiplied on the right. □

EXERCISE 2. Prove by an induction similar to that of Theorem 2.1:

THEOREM 2.2 *A system of n equations in n unknowns,*

$$\begin{cases} L_1 = a_{11}x_1 + a_{12}x_2 + \cdots + a_{1n}x_n = b_1 \\ L_2 = a_{21}x_1 + a_{22}x_2 + \cdots + a_{2n}x_n = b_2 \\ \quad\vdots \\ L_n = a_{n1}x_1 + a_{n2}x_2 + \cdots + a_{nn}x_n = b_n, \end{cases}$$

has a solution for any choice of $b_1, b_2, \ldots, b_n \in F$ *if and only if the system of homogeneous equations*

$$\begin{cases} L_1 = a_{11}x_1 + a_{12}x_2 + \cdots + a_{1n}x_n = 0 \\ L_2 = a_{21}x_1 + a_{22}x_2 + \cdots + a_{2n}x_n = 0 \\ \quad \vdots \\ L_n = a_{n1}x_1 + a_{n2}x_2 + \cdots + a_{nn}x_n = 0 \end{cases}$$

has only the trivial solution.

EXAMPLE. Interpolation of polynomials.

The coefficients of a polynomial

$$f(x) = c_0 + c_1 x + \cdots + c_{n-1}x^{n-1}$$

of degree $\le n - 1$ can be chosen to satisfy the linear equations in the c_i

$$\begin{cases} f(x_1) = \beta_1 \\ f(x_2) = \beta_2 \\ \quad \vdots \\ f(x_n) = \beta_n, \end{cases} \qquad x_i \ne x_j \text{ for } i \ne j$$

where $x_1, x_2, \ldots, x_n, \beta_1, \beta_2, \ldots, \beta_n$ are any preassigned numbers. This follows from the fact that the system

$$\begin{aligned} f(x_1) &= 0, \\ f(x_2) &= 0, \\ &\vdots \\ f(x_n) &= 0, \end{aligned}$$

of homogeneous linear equations has only the trivial solution since no polynomial of degree less than n can have n distinct roots.

2.2. Vector Spaces

A (left) *vector space* V over a field F is an additive commutative group. Its elements are called *vectors*. The identity of this group will be denoted by 0. There is an operation which assigns to any $a \in F$, $A \in V$, a product $aV = B \in V$. We ask that this operation satisfy the postulates:

(A) $$a(bA) = (ab)A,$$

(B) $$(a + b)A = aA + bA,$$

(C) $$a(A + B) = aA + aB,$$

(D) $$1 \cdot A = A.$$

The last postulate is not a consequence of the other three for we could define $aA = 0$ for all products and satisfy (A), (B), and (C).

EXERCISE 3. Show: (a) $a0 = 0$, (b) $0 \cdot a = 0$, (c) $-A = (-1) \cdot A$.

(a) We have

$$a0 + a0 = a(0 + 0) = a0.$$

Adding $-a0$ on both sides we obtain

$$a0 = 0.$$

(b) Similarly,

$$0A + 0A = (0 + 0)A = 0A.$$

Adding $-0A$ on both sides gives

$$0A = 0.$$

(c) From $(1 - 1)A = 0A = 0$ we have

$$1A + (-1)A = 0,$$

or $(-1)A$ is the inverse of A, i.e.,

$$(-1)A = -A.$$

From these results we can prove that $aA = 0$ implies either $a = 0$ or $A = 0$. If $a \neq 0$ then from $aA = 0$ we have

$$a^{-1}(aA) = 0 = (a^{-1}a)A = A.$$

Hence $A = 0$.

The n vectors $A_1, A_2, \ldots, A_n$ are said to be *linearly dependent* if there exist $x_1, x_2, \ldots, x_n \in F$ with not all $x_i = 0$ such that

$$x_1 A_1 + x_2 A_2 + \cdots + x_n A_n = 0. \tag{[1]}$$

Take $n = 1$. A vector A_1 is said to be linearly dependent if there exists an $x \neq 0$ in F such that $xA = 0$, i.e., if $A = 0$. If the vector is not zero it is independent. Assume that [1] holds for nontrivial x_i. Then we have, say, $x_n \neq 0$. It is possible to write

$$A_n = -x_n^{-1}x_1 A_1 - x_n^{-1}x_2 A_2 - \cdots - x_n^{-1}x_{n-1}A_{n-1}.$$

A sum of the form

$$c_1 A_1 + c_2 A_2 + \cdots + c_n A_n$$

is called a *linear combination* of the vectors $A_1, A_2, \ldots, A_n$. The statement that n vectors are linearly dependent is equivalent to the statement that one of them is a linear combination of the others.

The dimension of a vector space V is the maximum number of linearly independent vectors in V. If no such maximum exists the dimension of V is said to be infinite.

EXAMPLE. The polynomials form a vector space over the field of real numbers. In particular, the polynomials $1, x, \ldots, x^n$ are linearly independent. Clearly the dimension of the vector space of all polynomials is infinite.

The definition gives no hint of a way to obtain the dimension of any given vector space. In order to attack this problem we introduce

THEOREM 2.3 *Given n vectors $A_1, A_2, \dots, A_n \in V$ and if $B_1, B_2, \dots, B_m$ are $m > n$ linear combinations of the A_i, then the B_j are linearly dependent.*

PROOF: We are given the linear combinations

$$[1] \qquad \begin{cases} B_1 = a_{11}A_1 + a_{12}A_2 + \cdots + a_{1n}A_n \\ B_2 = a_{21}A_1 + a_{22}A_2 + \cdots + a_{2n}A_n \\ \vdots \\ B_m = a_{m1}A_1 + a_{m2}A_2 + \cdots + a_{mn}A_n. \end{cases}$$

For the proof of the theorem we must find $x_j \in F$ such that

$$[2] \qquad x_1B_1 + x_2B_2 + \cdots + x_mB_m = 0$$

where not all $x_j = 0$. Combining [1] and [2] we have

$$\sum_{j=1}^{m} x_jB_j = L_1A_1 + L_2A_2 + \cdots + L_nA_n = 0$$

where

$$[3] \qquad \begin{cases} L_1 = x_1a_{11} + x_2a_{21} + \cdots + x_ma_{m1} \\ L_2 = x_1a_{12} + x_2a_{22} + \cdots + x_ma_{m2} \\ \vdots \\ L_n = x_1a_{1n} + x_2a_{2n} + \cdots + x_ma_{mn}. \end{cases}$$

It suffices to find nontrivial x_j that make all $L_i = 0$. Since $m > n$, the system $L_i = 0$ of n equations in m unknowns has a nontrivial solution according to Theorem 2.1. It follows that there are x_i, not all of them zero, such that [2] holds and therefore the theorem is proved. □

COROLLARY *If V is a vector space in which all vectors are linear combinations of n given vectors, then the dimension of V is less than or equal to n.*

The vector space V is said to be *spanned* by the vectors $A_1, A_2, \dots, A_n \in V$ if every vector $B \in V$ is a linear combination of the A_i.

THEOREM 2.4 *If V is spanned by n linearly independent vectors, then the dimension N_V of V is precisely n.*

By the corollary to Theorem 2.3 we have $N_V \leq n$. But there exist n linearly independent vectors (e.g., $A_1, A_2, \dots, A_n$) in V and N_V is the maximum number of independent vectors in V. Consequently, $N_V \geq n$. Therefore $N_V = n$.

THEOREM 2.5 *If V is a vector space of finite dimension n, then there are n linearly independent vectors in V which span the space.*

PROOF: If n is the dimension of V, then V contains a set of n independent vectors; call them $A_1, A_2, \dots, A_n$. Let B be any vector of V. The $n+1$ vectors $B, A_1, A_2, \dots, A_n$ are linearly dependent since n is the maximum number of independent vectors. Thus there are $x_i \in F$, not all zero, such that

$$x_0B + x_1A_1 + \cdots + x_nA_n = 0.$$

It follows that $x_0 \neq 0$; for otherwise we would have $x_1A_1+x_2A_2+\cdots+x_nA_n=0$ where not all $x_1, x_2, \ldots, x_n$ are zero. The A_i, however, are linearly independent. Thus the possibility that $x_0 = 0$ is excluded. From $x_0 \neq 0$ we see at once that B is a linear combination of the A_i. We have proved, in fact, that *any* set of n linearly independent vectors in V spans the entire space. □

If W is a subspace of a finite-dimensional space V, then obviously the dimension of W is not greater than that of V. More precisely we have the

COROLLARY *If W is a subspace of V and the dimension of W is the same as the dimension of V, then $W = V$.*

PROOF: Let the dimension of W be n. Then there are n linearly independent vectors in W which span W. But these must also span V by last statement in the proof of the theorem. Therefore $W = V$. □

Given a field F and a number n, we can construct the vector space V_n over F consisting of all ordered n-tuples of elements of F. If

$$A = (a_1, a_2, \ldots, a_n), \quad a_i \in F,$$
$$B = (b_1, b_2, \ldots, b_n), \quad b_i \in F,$$

we define

$$A + B = (a_1 + b_1, a_2 + b_2, \ldots, a_n + b_n),$$
$$aA = (aa_1, aa_2, \ldots, aa_n).$$

EXERCISE 4. Verify that V_n satisfies the postulates for a vector space.

The dimension of the space V_n is easily seen to be n. From $0 = (0, 0, \ldots, 0)$ it follows that the n vectors

$$U_1 = (1, 0, 0, 0, \ldots, 0),$$
$$U_2 = (0, 1, 0, 0, \ldots, 0),$$
$$\vdots$$
$$U_n = (0, 0, \ldots, 0, 0, 1),$$

are linearly independent since

$$\sum_{i=1}^{n} c_iU_i = (c_1, c_2, \ldots, c_n)$$

is not zero unless all the c_i are zero. Furthermore, the n vectors span V_n since any vector $(c_1, c_2, \ldots, c_n) \in V_n$ can be written as a linear combination $\sum_{i=1}^{n} c_iU_i$. The result follows from Theorem 2.4.

EXERCISE 5. Show that any vector field of finite dimension n over F is isomorphic to V_n. By "V is isomorphic to V_n" we mean that V has essentially the same structure as V_n. In other words, to each element of one space there corresponds an element of the other which behaves in exactly the same manner under the operations among vectors. This concept will be dealt with later in a more precise manner.

In V_n, consider the equations

$$x_1 A_1 + x_2 A_2 + \cdots + x_n A_n = B. \quad [1]$$

Setting the components on one side equal to the corresponding components on the other, we obtain n linear equations in n unknowns as in Theorem 2.2. Equation [1] has a solution for all $B \in V_n$ if and only if the A_i are linearly independent and therefore span V_n. But this is equivalent to the assertion that the homogeneous equation

$$x_1 A_1 + x_2 A_2 + \cdots + x_n A_n = 0$$

has only the trivial solution, all $x_i = 0$. In terms of the components this is exactly the statement of Theorem 2.2.

CHAPTER 3

Polynomials. Factorization into Primes. Ideals.

In the following sections we shall devote considerable attention to the theory which has arisen from the attempts of algebraists to solve the general equation of nth degree

$$a_n x^n + a_{n-1}x^{n-1} + \cdots + a_0 = 0, \quad a_n \neq 0.$$

This is the central problem of algebra and it was principally to handle this problem that modern algebraic methods were developed. What is meant by a solution to such an equation? In analysis a solution is a method by which one can approximate as closely as one likes to a number that satisfies the equation. In algebra, however, the emphasis is on the nature and behavior of the solution. It is important, for example, to know whether or not an equation is solvable in radicals. In analysis, this question is not necessarily relevant.

3.1. Polynomials over a Field

Take for the domain of our discussion a commutative field F.[1] A *power series* over F is a sequence of elements of F which obeys certain rules of computation. A sequence of elements of F is simply a correspondence which associates with each nonnegative integer n exactly one element c_n of F. We denote a power series by

$$c_0 + c_1 x + c_2 x^2 + \cdots + c_n x^n + \cdots = \sum_{\nu=0}^{\infty} c_\nu x^\nu.$$

This notation is nothing more than a way of writing such a correspondence; it is not to be interpreted as a sum, no meaning is to be attached to the x's or their indices, and $c_\nu x^\nu$ is not to be considered as a product. A *polynomial* is a power series all of whose elements from a certain element on are zero. A polynomial could be defined as an ordered n-tuple of elements of F, but this would involve difficulties in framing rules of computation, which we avoid by handling power series as one may see by attempting to specialize the following rules to suit this definition:

The *sum* of two power series is defined by

$$\sum_{\nu=0}^{\infty} a_\nu x^\nu + \sum_{\nu=0}^{\infty} b_\nu x^\nu = \sum_{\nu=0}^{\infty} (a_\nu + b_\nu) x^\nu.$$

From this definition it follows at once that the power series form a commutative group under addition with the zero element $\sum_{\nu=0}^{\infty} 0 \cdot x^\nu$.

[1] All fields are assumed hereafter to be commutative unless the contrary is stated.

EXERCISE 1. Show that the polynomials are a subgroup of the group of power series under addition.

The *product* of two power series is defined by

$$\sum_{\nu=0}^{\infty} c_\nu x^\nu \sum_{\mu=0}^{\infty} d_\mu x^\nu = \sum_{n=0}^{\infty} e_n x^n$$

with

$$e_n = \sum_{\substack{\nu+\mu=n \\ \nu\mu\geq 0}} c_\nu d_\mu = \sum_{\nu=0}^{n} c_\nu d_{n-\nu}.$$

By proving the distributive law of multiplication over addition and the associative law of multiplication, we now show that the set of power series over F forms a ring.

The distributive law follows from the linearity of the product and from the distributive law for the field elements. We prove this in general. Let $\{a_n\}$ and $\{b_n\}$ be two sequences of elements in F. Define the product $\{a_n\} \cdot \{b_n\} = d_n$ to be linear in the a's and b's but otherwise arbitrary. Thus d_n is of the form $d_n = \Sigma\alpha_{ij}a_i b_j$ with $\alpha_{ij} \in F$. Consequently,

$$\{c_n\}[\{a_n\} + \{b_n\}] = \{d_n\} \text{ with } d_n = \sum \alpha_{ij} c_i (a_j + b_j) = \sum \alpha_{ij} c_i a_j + \alpha_{ij} c_i b_j$$

or

$$\{d_n\} = \{c_n\} \cdot \{a_n\} + \{c_n\} \cdot \{b_n\}.$$

EXAMPLES. The product of vectors in physics.

The scalar product is $a \cdot b = a_1 b_1 + a_2 b_2 + a_3 b_3$ and is therefore distributive. The vector product $a \times b$ has components of the form $\pm(a_i b_j - a_j b_i)$ and hence is distributive.

The associative law follows immediately from

$$\left(\sum a_\mu x^\mu \sum b_\nu x^\nu\right) \sum c_\rho x^\rho = \sum_{n=0}^{\infty}\left(\sum_{\mu+\nu=n} a_\mu b_\nu\right) x^n \sum c_\rho x^\rho$$
$$= \sum_{m=0}^{\infty}\left(\sum_{\mu+\nu+\rho=m} a_\mu b_\nu c_\rho\right) x^m.$$

Since the result is symmetrical, it is independent of the placing of the parentheses. This completes the proof that the power series forms a ring.

It is a simple matter to enlarge this ring to a field. First we note that the ring already has a multiplicative identity, namely

$$1 + 0 \cdot x + 0 \cdot x^2 + \cdots .$$

To obtain inverses with respect to multiplication, we have only to include all elements of the form

$$\sum_{n=-m}^{\infty} a_n x^n, \quad m > 0.$$

Elements of the form

$$\sum_{-\infty}^{+\infty} a_n x^n$$

cannot be included, for the nth coefficient of a product would be written

$$\sum_{\nu+\mu=n} a_\nu b_\mu = \sum_{\nu=-\infty}^{+\infty} a_\nu b_{n-\nu}.$$

This expression, however, is meaningless since the result of an infinite number of operations (in this case additions) is not defined.

EXERCISE 2. Prove that the product of two polynomials is a polynomial.

It follows from the closure of the set of polynomials with respect to multiplication and from the proof of Exercise 1 that the polynomials are a subring of the ring of power series. The multiplicative identity $1 + 0 \cdot x + 0 \cdot x^2 + \cdots$ is also a polynomial. This suggests

EXERCISE 3. Show how the ring of polynomials may be enlarged to a field.

A polynomial is completely described if its nonzero coefficients are given. This suggests the introduction of a finite notation for a polynomial which omits all terms with zero coefficients. We denote the polynomial

$$\sum_{k=0}^{\infty} a_k x^k \quad \text{with } a_k = 0 \text{ for } k > n \text{ by } \overline{\sum_{k=0}^{n} a_k x^k} \quad (a_k \neq 0)$$

where we adopt the convention that all terms with $a_k = 0$ are omitted from the barred symbol. In order to include the exceptional case we define

$$\bar{0} = 0 + 0 \cdot x + 0 \cdot x^2 + \cdots .$$

We have the particular cases:

$$\bar{a} = a + 0 \cdot x + 0 \cdot x^2 + \cdots ,$$
$$\bar{x} = 0 + 1 \cdot x + 0 \cdot x^2 + \cdots .$$

It is easy to show that computation with the barred symbols gives the same result as computation with the polynomials. For this purpose it is sufficient to prove

$$\overline{\sum_{k=0}^{n} a_k x^k} = \sum_{k=0}^{n} \bar{a}_k (\bar{x})^k.$$

We use induction on n. The statement is certainly true for $n = 0$ since

$$\sum_{k=0}^{0} a_n = \bar{a}_0.$$

If it is true for n, it must be true for $n + 1$.

Case 1. $a_{n+1} = 0$. The assertion is trivially true.

Case 2. $a_{n+1} \neq 0$. We have

$$\overline{\sum_{k=0}^{n+1} a_k x^k} = \overline{\sum_{k=0}^{n} a_k x^k} + \overline{a_{n+1}x^{n+1}} = \sum_{k=0}^{n+1} \overline{a_k}(\bar{x})^k + \overline{a_{n+1}}(\bar{x})^n(\bar{x})$$
$$= \sum_{k=0}^{n+1} \overline{a_k}(\bar{x})^k.$$

Since computation with the barred symbols is essentially the same as computation with the polynomials, the bar may be omitted without danger of confusion. Thus we have created new symbols for the polynomials for which the signs of addition and multiplication have meaning.

A polynomial $a_0 + a_1 x + \cdots + a_n x^n$ can be used to define the function $f(x)$ which assigns to any $c \in F$ and $f(c) \in F$ where $f(c) = a_0 + a_1 c + \cdots + a_n c^n$.

EXERCISE 4. If $f(x)$, $g(x)$ are polynomials and $c \in F$, show that

$$f(x) + g(x) = h(x) \Rightarrow f(c) + g(c) = h(c)$$

and

$$f(x) \cdot g(x) = h(x) \Rightarrow f(c) \cdot g(c) = h(c).$$

The *degree* of a polynomial is the highest index attached to a nonzero coefficient. If the polynomial is zero, it possesses no degree in the sense of this definition. To avoid the necessity of discussing special cases, however, the zero polynomial is assumed to have any negative degree.

EXERCISE 5. Given two nonzero polynomials $f(x)$ of degree m and $g(x)$ of degree n, show that $f(x)+g(x)$ has the degree $\max(m, n)$ if $m \neq n$ and $f(x) \cdot g(x)$ has the degree $m + n$.

EXERCISE 6. Prove the long division property for polynomials. That is, given two polynomials $f(x)$ and $g(x) \neq 0$, show that there are polynomials $q(x)$ and $r(x)$ such that

$$f(x) = q(x)g(x) + r(x) \qquad [1]$$

where the degree of $r(x)$ is less than the degree of $g(x)$.

PROOF: We consider two cases:

Case 1. There is a $q(x)$ with $f(x) = q(x)g(x)$. Consequently, $r(x) = 0$ and the statement is proved. In this case we say $f(x)$ is divisible by $g(x)$.

Case 2. No such $q(x)$ exists. In that event consider the set of polynomials of the form

$$f(x) - q(x)g(x). \qquad [2]$$

In this set there must be a polynomial of least degree; call it $r(x)$. The degree of $r(x)$ is less than the degree of $g(x)$. For suppose the degree of $g(x)$ is m and the degree of $r(x)$ is $n \geq m$, i.e.,

$$g(x) = a_0 + a_1 x + \cdots + a_m x^m, \quad a_m \neq 0,$$
$$r(x) = b_0 + b_1 x + \cdots + b_n x^n, \quad b_n \neq 0.$$

Then we may define a polynomial

$$r_1(x) = r(x) - (b_n/a_m)x^{n-m} g(x)$$

of degree $\leq n-1$. But from

$$r(x) = f(x) - q(x)g(x)$$

we have

$$r_1(x) = f(x) - [q(x) + (b_n/a_m)x^{n-m}]g(x)$$

or $r_1(x)$ is of the form [2] and has a degree less than that of $r(x)$. However, $r(x)$ was supposed to be the polynomial of type [2] of least degree. Consequently, the degree of $r(x)$ must be less than that of $g(x)$.

We observe first that the result of long division is unique. For suppose we have two representations

$$f(x) = q_1(x)g(x) + r_1(x)$$
$$f(x) = q_2(x)g(x) + r_2(x)$$

where the degrees of $r_1(x)$ and $r_2(x)$ are less than that of $g(x)$. This implies

$$[q_1(x) - q_2(x)]g(x) + [r_1(x) - r_2(x)] = 0.$$

Consequently, $q_1(x) - q_2(x) = 0$ since two polynomials of different degrees cannot be equal. Thus $r_1(x) - r_2(x) = 0$ and the proof is complete. □

An immediate consequence of the long division theorem is the familiar *remainder theorem.* Let $g(x) = x - a$ in [1]. Thus

$$f(x) = q(x)(x-a) + c.$$

Hence

$$f(a) = c \quad \text{or} \quad f(x) = q(x)(x-a) + f(a).$$

COROLLARY *The equation $f(x) = 0$ has the solution $x = a$ if and only if $f(x)$ is divisible by $x - a$.*

3.2. Factorization into Primes

A polynomial $f(x)$ over a field F is said to be *factored* if it can be written as the product of polynomials of positive degree:

$$f(x) = g(x) \cdot h(x) \cdots z(x).$$

The polynomials $g(x), h(x), \ldots, z(x)$ are called *factors* of $f(x)$. We shall consider two factorizations identical if one can be obtained from the other by rearranging the factors and multiplying each by some element of the field. If there are no two polynomials of positive degree which have the product $f(x)$, then $f(x)$ is said to be *irreducible* in F.

For the purpose of investigating the solutions of equations $f(x) = 0$, it is sufficient to consider irreducible polynomials. For, if $f(x) = g(x) \cdot h(x)$ and $f(a) = g(a) \cdot h(a) = 0$, then either $g(a) = 0$ or $h(a) = 0$. The polynomials have the important property that every polynomial possesses a "unique" factorization into irreducible polynomials, where by "unique" we mean that any two factorizations of the same polynomial into irreducible factors are identical. The similarity of this result and the theorem of unique factorization into primes for integers is quite striking. We are led to examine the properties common to the polynomials and the integers in order to uncover the general principle of which those are special cases. We note at once that the polynomials and the integers are both commutative rings, with an identity element for multiplication, for which the law $ab = 0 \Rightarrow a = 0$ or $b = 0$ holds. These conditions are not enough to guarantee a unique factorization into primes. As a counterexample, consider the numbers $a + b\sqrt{-3}$ where a and b are integers. Clearly this is a ring of the given type. Yet we have

$$4 = 2 \cdot 2 = (1 + -\sqrt{-3})(1 - \sqrt{-3}):$$

unique factorization does not hold in this ring.

EXERCISE 7. Prove that both factorizations of 4 in the ring of $a + b\sqrt{-3}$ are factorizations into primes.

Actually, it is the existence of long division which guarantees unique factorization into primes in the special cases of the polynomials and the integers. The long division theorem, however, involves the notion of "magnitude." In the case of polynomials it is the degree; in the case of integers it is the absolute value. This notion of magnitude is not necessary, however, as we shall show. What property of the ring is it that guarantees the unique factorization theorem and is implied by long division in these special cases?

3.3. Ideals

Consider a ring R. A subset of R is called an *ideal* $\mathfrak{A}$ if

(a) $\mathfrak{A}$ is a group with respect to addition

(b) $\left.\begin{array}{l} a \in \mathfrak{A} \\ b \in R \end{array}\right\} \Rightarrow ab \in \mathfrak{A}.$

THEOREM 3.1 *In the ring of integers there are no other ideals than those consisting of the multiples of a given integer and the set consisting of zero alone.*

PROOF: Let $\mathfrak{A}$ be an ideal in the ring of integers.

Case 1. $\mathfrak{A}$ consists of zero alone.

Case 2. There is a nonzero $a \in \mathfrak{A}$. If $a < 0$ then $-1 \cdot a = -a > 0$ and $-a \in \mathfrak{A}$. Thus if an ideal contains nonzero elements it also contains positive elements. From the set of positive integers in $\mathfrak{A}$ take the least and call it d. By (b) every multiple of d is an element of $\mathfrak{A}$. We prove that $\mathfrak{A}$ is precisely the set of multiples of d. Take any $a \in \mathfrak{A}$. By the division algorithm we have

$$a = qd + r, \quad 0 \le r < d.$$

But $a \in \mathfrak{A} \Rightarrow r = a - qd \in \mathfrak{A}$. Since d is the smallest positive integer in $\mathfrak{A}$ and $0 \leq r < d$, it follows that $r = 0$. Consequently, $a = qd$. Thus, any element of $\mathfrak{A}$ is a multiple of d. □

The same theorem holds for polynomials and its proof uses the division algorithm for polynomials in a similar way. This property of the integers does not hold for rings in which the unique factorization property does not hold, e.g., the ring of numbers $a + b\sqrt{-3}$ where a, b are integers.

EXERCISE 8. Show that the subset of elements for which $a + b$ is even form an ideal in the ring of $a + b\sqrt{-3}$. Prove that this ideal does not consist of the multiples of any one element. (See Exercise 7.)

We make the definition:

An ideal is called a *principal ideal* if it is the set of all multiples of a given element d of the ring.

Both for integers and for polynomials, where factorization is unique, the only ideals are principal. In one case where there is no unique factorization we have shown that this result does not hold. We shall prove that the unique factorization theorem is a consequence of the following postulates:

(1) Multiplication is commutative.
(2) There is a multiplicative identity $1 \in R$.
(3) $ab = 0 \Rightarrow$ either $a = 0$ or $b = 0$.
(4) Every ideal in R is principal.

3.4. Greatest Common Divisor

Let R be a ring satisfying postulates (1)–(4). Assume $a, b \in R$ and $ab \neq 0$. If there is a $c \in R$ such that $a \cdot c = b$, we say, variously, "b is a multiple of a," "b is divisible by a," and "a is a divisor of b."[2] We write $a|b$ (read: "a divides b"). The divisors of 1 are called the units of the ring.

EXAMPLES. In the ring of integers the units are ± 1. If R is the ring of polynomials over F its units are all $a \in F$, $a \neq 0$. The ring of Gaussian integers $a + bi$, where a and b are integers, possesses the units ± 1, $\pm i$. It is interesting that this is a principal ideal ring. The primes of the subring of ordinary integers are prime in this ring only if they are of the form $4n - 1$. All others are not prime, e.g., $5 = (1 + 2i)(1 - 2i)$. This is a consequence of the theorem that all primes of the form $4n + 1$ can be represented as the sum of two squares.

If $a|b$ and $b|c$, then $a|c$

$$\left.\begin{matrix} a|b \\ b|c \end{matrix}\right\} \Rightarrow \exists \alpha, \beta \in R \quad \text{such that} \quad \begin{cases} a\alpha = b \\ b\beta = c. \end{cases}$$

(Read: "there is (are)" or "there exists(s)" for "$\exists$.") Consequently, $a\alpha\beta = c$ or $a|c$.

[2]For rings containing elements $a \neq 0$, $b \neq 0$ such that $a \cdot b = 0$ is possible, we call a and b "divisors of zero."

If $a|b$, $a|c$ then $a|(b+c)$

$$\left.\begin{matrix} a|b \\ a|c \end{matrix}\right\} \Rightarrow \exists s, t \in R \quad \text{such that} \quad \begin{cases} as = b \\ at = c. \end{cases}$$

Therefore $a(s+t) = b+c$ or $a|(b+c)$.

If $a|b$ and $b|a$, the elements a and b have the same division properties, that is,

$$a|c \Leftrightarrow b|c \quad \text{and} \quad c|a \Leftrightarrow c|b.$$

PROOF:

$$\begin{matrix} b|a \\ a|c \end{matrix} \Rightarrow b|c. \qquad \begin{matrix} c|a \\ a|b \end{matrix} \Rightarrow c|b.$$

Since a and b appear symmetrically the proof is complete. □

If $a|b$ and $b|a$, then a and b are said to be *equivalent* (with respect to division). Two elements are equivalent if and only if they differ by a unit factor.

PROOF: Let $a, b \in R$ be equivalent, i.e., $a|b$ and $b|a$. This means that there are elements ε, η in R such that $a = \varepsilon b$ and $b = \eta a$. Therefore $a = \varepsilon\eta a$. This implies $a(\varepsilon\eta - 1) = 0$. Since $a \neq 0$, the use of postulate (3) gives $\varepsilon\eta = 1$. Thus $\varepsilon|1$, $\eta|1$. Conversely, if $\varepsilon\eta = 1$ and $b = \varepsilon a$, then b is equivalent to a. From $b = \varepsilon a$ we already have $a|b$. Multiplying by η we obtain $b = \eta\varepsilon a = a$ or $b|a$. □

Now suppose $a_1, a_2, \ldots, a_n \in R$. Consider the set

$$\mathfrak{A} = a_1R + a_2R + \cdots + a_nR^3$$

consisting of elements of the form

$$a_1x_1 = a_2x_2 + a_3x_3 + \cdots + a_nx_n, \tag{1}$$

where $x_1, x_2, \ldots, x_n \in R$. $\mathfrak{A}$ is an ideal. To prove this we have only to show that $\mathfrak{A}$ is an additive group (i.e., is closed under addition and subtraction) and is closed under multiplication by elements of R. Our result is immediate since

$$\sum_{i=1}^{n} a_ix_i \pm \sum_{i=1}^{n} a_iy_i = \sum_{i=1}^{n} a_i(x_i \pm y_i)$$

and

$$\left(\sum_{i=1}^{n} a_ix_i\right)z = \sum_{i=1}^{n} a_i(x_iz)$$

where $x_i, y_i, z \in R$. $\mathfrak{A}$ is a principal ideal by postulate (4) applied to R; therefore $\mathfrak{A}$ consists of the multiples of a single element d. We now write

$$\mathfrak{A} = dR = a_1R + a_2R + \cdots + a_nR;$$

that is,

$$a \subseteq \mathfrak{A} \Rightarrow \begin{cases} a \text{ is a multiple of } d, \text{ and} \\ a \text{ is expressible in the form [1].} \end{cases}$$

---

[3]The *sum* of two sets denoted by $S + T$ is the set of elements $s + t$ where $s \in S$, $t \in T$. The union (or logical sum) of the two sets is denoted differently by $S \cup T$.

Furthermore, $d \in \mathfrak{A}$ since, by postulate (2), $1 \in R$ and hence $1 \cdot d \in \mathfrak{A}$. Also $a_1, a_2, \dots, a_n \in \mathfrak{A}$; for we may take, say, $x_1 = 1$, $x_i = 0$ $(i > 1)$ in [1] above. Consequently, there are $x_1, x_2, \dots, x_n \in R$ such that

$$d = a_1x_1 + a_2x_2 + \cdots + a_nx_n$$

where $d|a_i$ $(i = 1, 2, \dots, n)$. Thus d is called a *common divisor* of the a_i.

Let δ be any common divisor of the a_i, i.e., $\delta|a_1$, $\delta|a_2, \dots, \delta|a_n$. It follows for any choice of the x_i that

$$\delta|a_1x_1 + a_2x_2 + \cdots + a_nx_n.$$

Hence δ is a divisor of all elements of $\mathfrak{A}$. Consequently, $\delta|d$. Conversely, since $d|a_i$ $(i = 1, 2, \dots, n)$, $\delta|d \Rightarrow \delta|a_i$. Thus, the common divisors of a_i and the common divisors of d are the same. Any element having this property is called a *greatest common divisor* of the a_i. The greatest common divisors of the a_i are equivalent under division. For if d and d' are greatest common divisors of the a_i we have $d|d'$ and $d'|d$. For this reason any greatest common divisor of the a_i is called *the* greatest common divisor. Equivalent elements will not usually be distinguished; there is no danger of confusion since the behavior of an element with respect to division is exactly the same as that of any of its equivalents.

A *linear diophantine equation* in R is an equation of the form

$$a_1x_1 + a_2x_2 + \cdots + a_nx_n = b.$$

Such an equation can have a solution if and only if b is a multiple of d, the greatest common divisor of the a_i. This is a direct consequence of $dR = a_1R + a_2R + \cdots + a_nR$.

EXAMPLE. The equation

$$32x + 74y - 18z = b$$

obviously has no solution in integers if b is odd. On the other hand it has solutions for all even b.

The elements $a_1, a_2, \dots, a_n$ of R are said to be *relatively prime* if 1 is their greatest common divisor. Thus, the integers 6, 10, 15 are said to be relatively prime. If the a_i are relatively prime the diophantine equation

$$a_1x + a_2x_2 + \cdots + a_nx_n = 1$$

has a solution. An element p is said to be *prime* if it has no divisors other than itself and 1 and if it does not divide 1. According to this definition the element 1 is *not* a prime.

THEOREM 3.2 *If a prime $p \in R$ divides a product ab, then it divides at least one of the factors, i.e.,*

$$\left.\begin{matrix} p|ab \\ p\nmid a \end{matrix}\right\} \Rightarrow p|b.$$

(Read: "p does not divide a" for $p\nmid a$.)

EXAMPLES. This theorem is true in general only for principal ideal rings. Consider, e.g., $2 \cdot 2 = (1+\sqrt{-3})(1-\sqrt{-3})$ in the ring of numbers $a + b\sqrt{-3}$ where a and b are integers. Again, in the ring consisting of the even integers 6 is prime and 18 is prime, yet we have $6 \cdot 6 = 2 \cdot 18$.

PROOF: The greatest common divisor of p and a is 1 since $p \nmid a$ and the divisors of p are only p and 1. Thus p and a are relatively prime and the equation $px + ay = 1$ has a solution $x, y \in R$. Multiplying both sides of the equation by b, we obtain

$$b = pbx + aby.$$

Since $p|ab$, the right side is divisible by p. Hence $p|b$. □

COROLLARY *If $p|a_1a_2\cdots a_n$ then p divides at least one if the a_i.*

PROOF: This theorem is true for $n = 1$. We use induction. Assume the theorem is true for n and suppose

$$p|a_1a_2a_3\cdots a_na_{n+1}.$$

By Theorem 3.2 either $p|a_{n+1}$ or $p|a_1a_2\cdots a_n$. In the former case the theorem is proved. In the latter case the theorem follows by the induction assumption. □

Suppose an element possesses a factorization into primes. Two such factorizations are said to be identical if the primes of one can be paired off with equivalent primes in the other. Thus identical factorizations are the same except for order and multiplication by the units. If all possible factorizations of an element are identical, the element is said to possess a *unique* factorization into primes.

THEOREM 3.3 *If an element possesses a factorization into primes, the factorization is unique.*

PROOF: Assume two factorizations

$$p_1p_2\cdots p_r = q_1q_2\cdots q_s.$$

First we have $r = s$. For each p_i divides a q_k and no q_k possesses more than one p_i as a divisor. Therefore $s \geq r$. Similarly $r \geq s$. Therefore $r = s$. Now $p_1|q_1q_2\cdots q_r$. By the corollary to Theorem 3.2, p_1 is a divisor of one of the q_i; $p_1|q_1$ say. Since q_1 has only itself and 1 as divisors and $p_1 \nmid 1$ it follows that p_1 and q_1 are equivalent, i.e., $p = \varepsilon q_1$. Consequently, we may write

$$\varepsilon q_1 p_2 \cdots p_r = q_1q_2\cdots q_r$$

or

$$(\varepsilon p_2)\cdots p_r = q_2\cdots q_r.$$

The theorem follows by induction. □

It is conceivable that there are elements which possess no decomposition into primes. In other words, an element might be factored in such a way that nonprimes are included in the factorization no matter how far the process is carried. For integers and polynomials there is no such danger since the number of elements in the product is limited by the "magnitude" of the element being factored. However, the result is true in general and therefore every element of R possesses a unique factorization into primes.

LEMMA *Let $a_1, a_2, \ldots \in R$ be a sequence of nonzero elements such that $a_{i+1}|a_i$ for all i. Then all the a_i from a given element on are equivalent.*

PROOF: Let $\mathfrak{A}$ be the set of all multiples of the a_i. $\mathfrak{A}$ is an ideal; for take any $a, b \in \mathfrak{A}$. We have

$$a \in \mathfrak{A} \Leftrightarrow a = a_i c,$$
$$b \in \mathfrak{A} \Leftrightarrow b = a_j d.$$

Assume $i \geq j$, say. Then $a_i|a_j$. Therefore $\exists s \in R$ with $b = a_i s$. Hence $a \pm b = a_i(c \pm s)$; $\mathfrak{A}$ is closed with respect to multiplication. Furthermore, $\mathfrak{A}$ is closed with respect to multiplication by elements of R since for $r \in R$, $a \cdot r = a_i(cr)$. $\mathfrak{A}$ is a principal ideal by postulate (4) and hence there is a $d \in R$ such that $\mathfrak{A} = dR$. Thus $d \cdot 1 \in \mathfrak{A}$ and d is in $\mathfrak{A}$. Therefore there is an a_n which divides d. Consequently,

$$a_n, a_{n+1}, a_{n+2}, \ldots |d.$$

But $a_i|a_i \Rightarrow a_i \in \mathfrak{A} \Rightarrow a_i \in dR$. Hence $d|a_n, a_{n+1}, a_{n+2}, \ldots$. We have proved that all the a_i for $i \geq n$ are equivalent to d. □

THEOREM 3.4 *Every $a \in R$ is either zero, a unit, a prime, or a product of primes.*

PROOF: Suppose a is none of these, i.e., $a \neq 0$, $a \nmid 1$, and a is neither prime nor a product of primes. Since a is not prime it can be expressed as a product $bc = a$ where neither b nor c is equivalent to a. Clearly $b \neq 0$, $c \neq 0$. If c and b were each either a unit, a prime, or a product of primes, then a would be in one of these categories. This possibility is ruled out. It follows that one of the divisors, say b, has the same property as a. But this reasoning could be carried out indefinitely to give a sequence of elements satisfying the hypothesis of the lemma but for which the terms do not eventually become equivalent. This indirect proof establishes the theorem. □

We have proved that every element can be factored uniquely into primes. Suppose a has the factorization

$$a = p_1 p_2 \cdots p_r$$

where the p_i may be the same or distinct. It is possible that the same prime and its equivalents may appear more than once in this expression. If all equivalent elements are taken together we may write

$$a = \varepsilon p_1^{\nu_1} p_2^{\nu_2} \cdots p_s^{\nu_s}, \quad \nu_i > 0,$$

where the p_i are now essentially distinct, i.e., $p_i \nmid p_j$, $i \neq j$. Clearly $\nu_1 + \nu_2 + \cdots + \nu_s = r$. Any element of the form

$$d = p_1^{\mu_1} p_2^{\mu_2} \cdots p_s^{\mu_s}, \quad 0 \leq \mu_i \leq \nu_i,$$

is obviously a divisor of a. Conversely, if $d|a$ it must be of this form since the factorization of d can contain no prime to a higher degree than its degree in a. We may now find an expression for the greatest common divisor of two elements in terms of their factorizations. Suppose $a, b \in R$ with the factorizations

$$a = p_1^{\nu_1} p_2^{\nu_2} \cdots p_r^{\nu_r}, \quad b = p_1^{\mu_1} p_2^{\mu_2} \cdots p_r^{\mu_r},$$

in which we understand that μ_k, ν_k are not both zero. Thus every prime which appears in either factorization appears in both, if only nominally. The greatest common divisor of a and b is then

$$d = p_1^{\alpha_1} p_2^{\alpha_2} \cdots p_r^{\alpha_r}$$

where $\alpha_i = \min(\nu_i, \mu_i)$. In a similar way we write the least common multiple

$$D = p_1^{\beta_1} p_2^{\beta_2} \cdots p_r^{\beta_r}$$

where $\beta_i = \max(\nu_i, \mu_i)$.

CHAPTER 4

Solution of the General Equation of nth Degree. Residue Classes. Extension Fields. Isomorphisms.

4.1. Congruence

Consider the notation $a = b$. The sign of equality means that a and b are merely two ways of writing the same element. In other words, the symbols a and b are interchangeable in any discussion. We have already considered relations which are somewhat like equality in this respect. For example, in the preceding section "a is equivalent to b" means that a and b are interchangeable in any discussion of divisibility properties. Let us investigate relations of this kind in somewhat greater generality.

Assume we are given a set S of elements $a, b, c, \ldots$. A relation

$$a \equiv b$$

(read: "a congruent b") between two elements of S is called *congruence* (or equivalence or similarity) if it satisfies the postulates

(A)	$a \equiv a$	(reflexivity),
(B)	$a \equiv b \Rightarrow b \equiv a$	(symmetry),
(C)	$a \equiv b,\ b \equiv c \Rightarrow a \equiv c$	(transitivity).

EXAMPLES. A relation need not satisfy any of the postulates. For instance, let S be the set of human beings with the relation "a loves b." Every day "a loves a" is violated by some suicide. Furthermore, "a loves b" is nonsymmetric as any reader of novels can tell. True, an argument can be made in favor of the transitivity of this relation under the principle of "Love me, love my dog"—but the logic is dubious. For a set of people gathered in a pitch dark room at a seance we have the relation "a can see b" which vacuously satisfies the last two postulates, but not the first. A more orthodox example is the relationship "a approximates b" among the real numbers. If we understand this to mean that the difference between a and b lies within some given limit of error, we see then this relation is reflexive and symmetric but not transitive. A relation which violates only the symmetric law is "$a \leq b$" in the set of integers. We have shown by the last three examples that the postulates of a congruence relation are independent; i.e., no postulate can be derived logically from the other two.

By means of the congruence relation the elements of S can be classified into nonoverlapping "species." For define S_a as the set of all $s \in S$ such that $s \equiv a$. If S_a and S_b overlap at all they are completely identical. For suppose $\exists c \in S$ such that

$c \in S_a$, $c \in S_b$. Then $c \equiv a$, $c \equiv b$. By postulates (B) and (C), $a \equiv b$. If $d \in S_a$ then $d \equiv a$ and hence $d \equiv b$. Conversely, any element in S_b is in S_a. Thus the classes S_a do not overlap. Conversely, a covering of S by nonoverlapping subsets furnishes a congruence relation; namely, $a \equiv b$ if a and b are in the same subset.

Let R be a commutative ring and assume a congruence relation in R which satisfies postulates (A), (B), (C) and is preserved by the operations in the ring, i.e.,

$$\text{(D)} \qquad a \equiv c,\ b \equiv d \Rightarrow \begin{array}{l} a + b \equiv c + d \\ a \cdot d \equiv c \cdot d. \end{array}$$

EXAMPLE. Let R be the set of integers with the added relation: Two integers are congruent if their difference is even. This is clearly a congruence of the above type. By means of this congruence we divide R into two classes, the even numbers and the odd numbers. Note that these two classes are the elements of the ring whose operations are defined by the tables at the beginning of Chapter 2.

A congruence relation which satisfies (D) will always define a ring in the same manner as in the above example. To prove this result, we first define

$$S_{a+b} = S_a + S_b$$

and

$$S_{a \cdot b} = S_a \cdot S_b;$$

i.e., S_{a+b} is the set of elements $c + d$ where $c \equiv a$, $d \equiv b$.

EXERCISE 1. Show that this definition is consistent with the former definition of S_{a+b} as the set of all elements congruent to $a + b$.

LEMMA *The sets S_x, $x \in R$, form a commutative ring.*

PROOF: First we show that they constitute a commutative additive group. Closure is obvious. The associative law is trivial:

$$(S_a + S_b) + S_c = S_{(a+b)+c} = S_{a+(b+c)} = S_a + (S_b + S_c).$$

There is an identity element:

$$S_a + S_0 = S_a.$$

To each element there is an inverse:

$$S_a + S_{-a} = S_0.$$

The commutative law is obvious. Next, for multiplication, the distributive law holds by

$$S_a(S_b + S_c) = S_{a(b+c)} = S_{ab+ac} = S_{ab} + S_{ac} = S_a \cdot S_b + S_a \cdot S_c.$$

The associative and commutative laws for multiplication clearly hold. □

Consider the set $\mathfrak{A} = S_0$ of all $a \equiv 0$.

(1) $\mathfrak{A}$ is closed with respect to addition and subtraction. For

$$a, b \in \mathfrak{A} \Rightarrow a \equiv 0,\ b \equiv 0 \Rightarrow a + b \equiv 0.$$

Furthermore, we have $a \equiv b$, but

$$a \equiv b,\ -b \equiv -b \Rightarrow a - b \equiv 0.$$

Hence $a, b \in \mathfrak{A} \Rightarrow a \pm b \in \mathfrak{A}$.

(2) $\mathfrak{A}$ is closed with respect to multiplication by elements of R

$$\left.\begin{array}{l} a \in \mathfrak{A} \\ b \in R \end{array}\right\} \Rightarrow \left.\begin{array}{l} a \equiv 0 \\ b \equiv b \end{array}\right\} \Rightarrow ab \equiv 0 \Rightarrow ab \in \mathfrak{A}.$$

Consequently, $\mathfrak{A}$ is an ideal. Thus we have shown that a congruence relation which is preserved under the operations in R defines an ideal, the set of elements $a \equiv 0$.

Conversely, let $\mathfrak{A}$ be an ideal in R. Use $\mathfrak{A}$ to define a new congruence relation: $a \equiv b$ means $a - b \in \mathfrak{A}$.

EXAMPLE. If $\mathfrak{A}$ is not an ideal, a congruence cannot be defined in this manner. Suppose, for example, that $\mathfrak{A}$ is the set of odd integers; then, by this rule $a \not\equiv a$. That the new relation is a congruence follows from the single fact that $\mathfrak{A}$ is an additive group.

(1) We have

$$0 \in \mathfrak{A} \Rightarrow a \equiv a.$$

(2) $a \equiv b \Rightarrow a - b \in \mathfrak{A} \Rightarrow -(a - b) \in \mathfrak{A}$. Hence

$$b - a \in \mathfrak{A} \quad \text{or} \quad b \equiv a.$$

(3)

$$\left.\begin{array}{l} a \equiv b \\ b \equiv c \end{array}\right\} \Rightarrow a - b,\ b - c \in \mathfrak{A}.$$

Therefore $a - c \in \mathfrak{A}$ or $a \equiv c$. We show further that this congruence satisfies (D).

(4)

$$\left.\begin{array}{l} a \equiv b \\ c \equiv d \end{array}\right\} \Rightarrow a - b,\ c - d \in \mathfrak{A}.$$

Using the group property we have

$$a + c - (b + d) \in \mathfrak{A} \quad \text{or} \quad a + c \equiv b + d.$$

Using closure under multiplication by elements of R, we obtain

$$(a - b)c,\ b(c - d) \in \mathfrak{A} \Rightarrow (a - b)c + b(c - d) \in \mathfrak{A}.$$

Consequently,

$$ac - bd \in \mathfrak{A} \quad \text{or} \quad ac \equiv bd.$$

By means of this new congruence relation we may now define an ideal S_0, the set of all elements $a \equiv 0$. But

$$a \in S_0 \Leftrightarrow a \equiv 0 \Leftrightarrow a - 0 \equiv a \in \mathfrak{A}.$$

Clearly, the specification of a congruence relation of this type and the specification of an ideal are completely equivalent.

The congruence defined in R by means of the ideal $\mathfrak{A}$ is denoted by

$$a \equiv b \pmod{\mathfrak{A}}.$$

The classes $S_a, S_b, \ldots$ are called the *residue classes* (mod $\mathfrak{A}$). In a principal ideal ring, $\mathfrak{A}$ consists of the multiples of one element d. In that case we use the notation (mod d) instead of (mod $\mathfrak{A}$).

EXAMPLES. Consider the congruence defined in the set of integers by the ideal consisting of the multiples of 7

$$a \equiv b \pmod 7 \Rightarrow 7 | (a - b).$$

Thus $a \equiv b$ means $a = 7m + b$: An integer is congruent to its remainder after division by 7. The ring of integers is split thereby into the seven residue classes $S_0, S_1, \ldots, S_6$. These classes are the elements of a commutative ring. We have, for example, $S_2 + S_4 = S_6$, $S_2 \cdot S_4 = S_1$, $S_3 + S_5 = S_1$. It is convenient to omit the S's and denote the elements of the ring by the subscripts $0, 1, \ldots, 6$ alone. The ring contains a multiplicative identity 1. We further note that all nonzero elements have inverses:

element	0	1	2	3	4	5	6
inverse		1	4	5	2	3	6

Thus the residue classes (mod 7) form a field. Linear equations may be solved in the usual way; if $3x = 4$, then $x = 3^{-1}4 = 5 \cdot 4 = 6$. Quadratic equations can be solved by completing the square; thus

$$x^2 + x + 1 = 0 = \left(x + \frac{1}{2}\right)^2 + \frac{3}{4} = (x+4)^2 - 1 = 0,$$

and we obtain the solutions

$$x = -4 \pm 1 \quad \text{or} \quad x = 4, \ x = 2.$$

Not all equations of degree higher than one have solutions; e.g., consider $x^2 - 3 = 0$.

If an integer m is not prime, then the ring of integers (mod m) is not a field. For we have divisors $a \cdot b = m$ with $a, b \neq m$ and hence we have divisors of zero.

EXAMPLE. The integers (mod 12) do not form a field since we have $3 \cdot 4 = 0$, for example. Thus there are divisors of zero; these elements do not have inverses.

THEOREM 4.1 *Let R be a principal ideal ring as defined by postulates* (1)–(4) *in Chapter* 3. *If $p \in R$ is a prime, the ring of the residue classes* (mod p) *is a field.*

PROOF: The ring R (mod p) has a unit element (i.e., multiplicative identity) S_1. It is sufficient to show closure under division; $S_a \neq S_0$ implies that S_a has an inverse. If $S_a \neq S_0$ then $a \in S_a \Rightarrow a \not\equiv 0 \pmod{p}$. Hence $p \nmid a$. Since the only divisors of p are p and 1, the greatest common divisor of p and a is 1. Therefore we can find $x, y \in R$ such that $ax + py = 1$. It follows that $ax \equiv 1 \pmod{p}$. Consequently, $S_a S_x = S_1$. □

4.2. Extension Fields

Consider the field F of integers (mod 7) and construct a table of their squares

x	0	1	2	3	4	5	6
x^2	0	1	4	2	2	4	1

The equation $x^2 = 3$ has no solution in this field. What, then, does it mean to solve the equation? In order to answer our question, consider the field of real numbers and the equation $x^2 + 1 = 0$, which has no solution in this field. In order to solve $x^2 + 1 = 0$ we construct the larger field of numbers $a + bi$, with a, b real and $i^2 = 1$. The same construction which leads from the real numbers to the complex numbers allows us to "solve" the most general equation with real coefficients. In our example, the equation $x^2 - 3 = 0$ has a solution in the field of numbers $a + b\sqrt{3}$, where a and b are integers (mod 7).

EXERCISE 2. Verify that the set of numbers $a + b\sqrt{3}$, where a, b are integers (mod 7), actually is a field.

An *extension field* of a field F is a field E such that $E \supset F$ and the operations for which E is a field coincide in F with those already defined. F is then called a *ground field* with respect to E. To solve an equation $f(x) = 0$ where $f(x)$ is a polynomial over F means to find an extension field of F which contains an element α such that $f(\alpha) = 0$. The element α is then called a *root* of $f(x)$. Let $p(x)$ be an irreducible polynomial over F. If $p(x)$ has degree 1 it has a root in F. If the degree of $p(x)$ is higher than 1, $p(x)$ cannot have a root in F; since

$$a \in F, \quad p(a) = 0 \Rightarrow (x - a) | p(x),$$

so that $p(x)$ would be reducible.

THEOREM 4.2 *Denote by $F(x)$ the ring of polynomials over F. If $p(x) \in F(x)$ is irreducible, then there is an extension field E of F which is "essentially" the ring of residue classes $\overline{E} = F(x) \pmod{p(x)}$. We further assert that $p(x)$ has a root in E.*

PROOF: The ring $F(x) (\text{mod } p(x))$ is a field $\overline{E}$ by Theorem 4.1. If $\phi(x) \in F(x)$ we may write $\phi(x) = c_0 + c_1 x + \cdots + c_n x^n$. The residue class $S_{\phi(x)} \in \overline{E}$ may be written

$$S_{c_0 + c_1 x + \cdots + c_n x^n} = S_{c_0} + S_{c_1} S_x + \cdots + S_{c_n} S_x^n.$$

The residue classes can therefore be described in terms of only two types, S_a where $a \in F$, and the class S_x. All others can be obtained from these by additions and multiplications.

Let us consider $\overline{F} = F \pmod{p(x)}$, the set of the S_a, $a \in F$. $S_a = S_b$ means $a \equiv b \pmod{p(x)}$ or $p(x)|(a-b)$. It follows that $a - b = 0$ or $a = b$. Thus every element of $\overline{F}$ contains only one element of F. Furthermore, every element of F belongs to an element of $\overline{F}$. The elements of $\overline{F}$ can be paired off with the elements of F. Moreover, we have

$$S_a + S_b = S_{a+b}, \quad S_a \cdot S_b = S_{a \cdot b},$$

so that the computation with the elements of $\overline{F}$ is in no way different from computation with the elements of F. We have shown that F is isomorphic to $\overline{F}$.[1] Now consider the equation

$$p(x) = a_0 + a_1 x + \cdots + a_n x^n = 0.$$

The corresponding equation with coefficients in $\overline{F}$ is

$$S_{a_0} + S_{a_1} X + \cdots + S_{a_n} X^n = S_0.$$

This possesses the solution $X = S_x$ in $\overline{E}$, for we have

$$S_{a_0} + S_{a_1} S_x + \cdots + S_{a_n} S_x^n = S_{a_0 + a_1 x + \cdots + a_n x^n} = S_{p(x)} = S_0.$$

We have obtained an extension field $\overline{E}$ of F, and shown that it is isomorphic to F. In order to obtain an extension field E of F we have only to replace those elements of $\overline{E}$ which are in $\overline{F}$ by the corresponding elements of F. We define addition and multiplication for E as follows: whenever S_a occurs in the tables of addition or multiplication for $\overline{E}$ replace it by a. For example, $S_a + S_x = S_{a+x}$ in $\overline{E}$ leads to $a + S_x = S_{a+x}$ in E. A field is defined when we are given its elements and rules of operation. Even though the elements of E have a mixed nature—some of them are classes of polynomials, others elements of F—it is nonetheless a perfectly good field. By constructing this field E we have proved the theorem. $\square$

The introduction of complex numbers into analysis proceeds in the manner of this theorem as an extension of the real numbers that contains a root of $x^2 + 1$. Long division by $x^2 + 1$ is so simple that one may immediately write down the residue classes.

EXERCISE 3. Solve the equation $x^2 + 1 = 0$ by extending the field of real numbers.

We now develop a more careful description of the elements of E. The most general residue class in E is $S_{\phi(X)}$, $\phi(x) \in F(x)$. Since $\partial[p(x)] = n$ (read: "the degree of $p(x)$" for "$\partial[p(x)]$"), we may assume $\partial[\phi(x)] < n$. For we can express any polynomial $\phi(x)$ in the form

$$\phi(x) = q(x)p(x) + r(x), \quad \partial[r(x)] < n.$$

Hence $\phi(x) \equiv r(x) \pmod{p(x)}$ or $S_{\phi(x)} = S_{r(x)}$. On the other hand, suppose $\phi(x), \psi(x) \in F(x)$ with

$$\partial[\phi(x)], \quad \partial[\psi(x)] < n.$$

[1] See Section 4.3.

If $S_{\phi(x)} = S_{\psi(x)}$, it follows that $p(x)|[\phi(x) - \psi(x)]$ and hence $\phi(x) - \psi(x) = 0$ or $\phi(x) = \psi(x)$.

EXAMPLE. $a + bi = c + di \Rightarrow a = c,\ b = d$. The sum $S_{\phi(x)} + S_{\psi(x)} = S_{\phi(x)+\psi(x)}$ is at once in the prescribed form since $\partial[\phi(x) + \psi(x)] < n$.

For the product, however, the result is not so simple. We may write

$$\phi(x)\psi(x) = q(x)p(x) + r(x).$$

This yields

$$S_{\phi(x)} \cdot S_{\psi(x)} = S_{r(x)}.$$

Only in those cases where $\partial[\phi(x) \cdot \psi(x)] < n$ is $S_{\phi(x)\cdot\psi(x)}$ immediately an element of the prescribed form.

We have shown that the elements of E are expressible in the form

$$S_{c_0+c_1x+\cdots+c_{n-1}x^{n-1}} = c_0 + c_1 S_x + \cdots + c_{n-1} S_x^{n-1}.$$

Two such elements are equal if and only if corresponding coefficients are equal.

EXAMPLE. Let F be the field of integers (mod 7). The equation $p(x) = x^3 - x + 2 = 0$ has no solution in F. Consequently, since $p(x)$ is of third degree it is irreducible; for any factorization of $p(x)$ would have to contain a linear factor. (On the other hand a fourth-degree polynomial might have two quadratic factors.) The extension field E consists of all elements $a + b\alpha + c\alpha^2$ where $a, b, c \in F$ and

$$p(\alpha) = \alpha^3 - \alpha - 2 = 0.$$

We have $\alpha^3 = \alpha + 2$, $\alpha^4 = \alpha^2 + 2\alpha$, $\alpha^5 = \alpha^3 + 2\alpha^2 = 2\alpha^2 + \alpha + 2, \ldots$. In this manner any power of α—and hence any polynomial in α—can be reduced to one of degree not greater than two.

Let F be a field, E an extension of F. Suppose $\alpha \in E$. We distinguish two cases:

Case 1. There is no nonzero polynomial over F which has α as a root.

EXAMPLE. Take F to be the field of rational numbers. The real number $e = 2.718\ldots$ is the root of no polynomial with rational coefficients. If α is an element of this type it is said to be *transcendental* with respect to F.

Case 2. If α is not transcendental, there is a polynomial $f(x) \in F(x)$ such that $f(\alpha) = 0$. We say, then, that α is algebraic with respect to F. Among all the polynomials which have the root α, there is one of least positive degree. Denote this by $p(x)$. Since $p(\alpha) = 0$ any multiple of $p(x)$ is also a polynomial with the root α. Conversely, suppose $f(x) \in F(x)$ and $f(\alpha) = 0$. We can find $q(x)$, $r(x) \in F(x)$ such that

$$f(x) = q(x)p(x) + r(x),$$

where $\partial[r(x)] < \partial[p(x)]$. Substituting α for x, we obtain

$$f(\alpha) = q(\alpha)p(\alpha) + r(\alpha),$$

whence $r(\alpha) = 0$. But $p(x)$ was assumed to be a polynomial of lowest positive degree which has the root α. Therefore $r(x) = 0$ or $f(x) = q(x)p(x)$. We have proved

LEMMA 4.3 *The polynomials for which α is a root are the multiples of the polynomial $p(x)$ of lowest degree.*

EXAMPLE. The number $\sqrt{2}$ is a root of the quadratic polynomial $x^2 - 2$ over the field of rational numbers. It cannot be a root of a lower-degree polynomial since it is irrational. Consequently, any polynomial which has $\sqrt{2}$ as a root must be a multiple of $x^2 - 2$.

LEMMA 4.4 *The polynomial $p(x)$ is irreducible in F. For otherwise $p(x) = a(x) \cdot b(x)$ where $\partial[a(x)], \partial[b(x)] > 0$.*

It follows that $p(\alpha) = a(\alpha) \cdot b(\alpha) = 0$, whence either $a(\alpha) = 0$ or $b(\alpha) = 0$. This contradicts the assumption that $p(x)$ is a polynomial of least degree.

LEMMA 4.5 *The only irreducible polynomials which possess the root α are the polynomials $c \cdot p(x)$, $c \in F$.*

Hence any polynomial of lowest degree is equivalent to $p(x)$. Therefore, according to the convention of the last chapter, we say $p(x)$ is *the* polynomial of lowest degree.

Suppose F is a ground field and $p(x)$ an irreducible polynomial over F. Let E be an extension of F which contains a root α of $p(x)$. By Lemma 4.5 $p(x)$ is the polynomial of least degree for which $p(\alpha) = 0$. However, E may certainly include elements which are not necessary for the solution of $p(x) = 0$. For example, if F is the field of rational numbers and our problem is to solve $x^2 - 2 = 0$, it is not necessary to extend F as far as the real numbers. What is the smallest field between F and E which contains α?

The required field certainly must contain every element of the form

$$\phi(\alpha) = c_0 + c_1\alpha + \cdots + c_n\alpha^n$$

where $\phi(x) \in F(x)$. Let us determine how these particular elements, the polynomials in α over F, equate, add, and multiply.

Suppose $\phi(\alpha) = \psi(\alpha)$. Then $\phi(x) = \psi(x)$ has the root α, whence

$$p(x)|[\phi(x) - \psi(x)]$$

(Lemma 4.3), or

$$\phi(x) \equiv \psi(x) \pmod{p(x)}.$$

Conversely, if $\phi(x) - \psi(x) = p(x)q(x)$, then

$$\phi(\alpha) - \psi(\alpha) = 0.$$

Thus we have proved

$$\phi(\alpha) = \psi(\alpha) \Leftrightarrow \phi(x) \equiv \psi(x) \pmod{p(x)}.$$

The rules for addition and multiplication of these elements are obviously the same as for polynomials. Thus we see that the set consisting of the elements $\phi(\alpha)$ is isomorphic to the set of the residue classes (mod $p(x)$) of the polynomials over F.

4.3. Isomorphism

The notion of isomorphism has already been touched upon here and there in the text. We have mentioned an "essential sameness" of two mathematical systems. It has been implied that two systems which are isomorphic differ in no important way; operations on the elements of one are "the same as" operations on the elements of the other. The purpose of this section is to replace this descriptive terminology by a precise formulation.

Implicit in the idea of the "essential sameness" of two sets is the knowledge that each element of one set has an "image" in the other. Specifically, consider two sets S and T. The set S is *mapped into* the set T if to each $s \in S$ there corresponds a $t \in T$, the *image* of s. The statements "S is mapped into T" and "t is the image of s" are denoted by

$$S \xrightarrow{\text{into}} T \quad \text{and} \quad s \to t,$$

respectively. A mapping is nothing more than a single-valued function with arguments in the set S and values in the set T. We could have written $t = f(s)$ instead of $s \to t$.

If every element of T is an image for some element of S, we say that S is mapped *onto* T and write $S \xrightarrow{\text{onto}} T$.

EXAMPLE. For the set S take a group G. Let the elements of T be the cosets of some subgroup $H \subset G$. By means of the mapping $f(x) = xH$, G is mapped onto the cosets of H. For an isomorphism we require more, as this example shows. Different elements of S should have distinct images. Thus if $s_1 \neq s_2$

$$\begin{matrix} s_1 \to t_1, \\ s_2 \to t_2, \end{matrix} \qquad t_1 \neq t_2.$$

From a mapping $S \xrightarrow{\text{onto}} T$ of this kind we can derive an inverse mapping $T \xrightarrow{\text{onto}} S$. For any $t \in T$ there is a single $s \in S$ such that $s \to t$. For the inverse mapping take $t \to s$. The mapping has furnished a method of pairing off the elements of S and T. In other words there is a one-to-one correspondence between the elements of S and the elements of T. In this case we say that the mapping is 1-1 (read: "one-to-one" for "1-1".)

EXAMPLES. The ordinary photographic image of a three-dimensional object does not provide a 1-1 mapping, if, say, the object is transparent. Take $p(x)$, an irreducible polynomial over a field F, and let α be one of its roots in some extension field $E \supset F$. For S take the set consisting of all the elements of E which derive from addition and multiplication of α with the elements of F; S is the set of all polynomials $\phi(\alpha)$ over F. For T take the set of residue classes $T_{\phi(x)}$ where $\psi(x) \in T_{\phi(x)}$ means $\psi(x) \equiv \phi(x) \pmod{p(x)}$. We have shown that

$$\phi(\alpha) = \psi(\alpha) \Rightarrow \phi(x) \equiv \psi(x) \pmod{p(x)}.$$

From this statement it is easy to see that there is a 1-1 correspondence between the elements of S and T, namely, $\phi(\alpha) \leftrightarrow T_{\phi(x)}$. Furthermore, we have

$$\phi(\alpha) + \psi(\alpha) \leftrightarrow T_{\phi(x)+\psi(x)} = T_{\phi(x)} + T_{\psi(x)},$$

and a similar result holds for multiplication. Thus the sum (or product) of the images of two elements is the image of the sum (or product) of the elements. This is what is meant by the essential sameness of two fields.

What do we mean when we say that two mathematical systems have the same structure? Before we answer this question it is necessary to specify what we mean by a mathematical system. A mathematical system is concerned with fundamental elements of various classes $S_1, S_2, \ldots$. Everything else is defined in terms of these elements. In analysis, the fundamental elements are real numbers; in geometry, points, lines, planes, ... For simplicity, let us assume that the fundamental elements are all of one class S. Relations are defined for the elements of S. A relation $R(x_1, x_2, \ldots, x_n)$ is a statement involving the elements $x_1, x_2, \ldots, x_n$. We do not mean to imply by this notation that the number of elements in a relation is finite. For example, the statement that a sequence of real numbers has a limit is a perfectly good relation.

To write a relation for specific elements does not mean that it is true.

EXAMPLES. If S is the set of integers and $R(x_1, x_2)$ means $x_1 = x_2$ then $R(5, 5)$ is true but $R(5, 7)$ is not. Label the vertices of a square in accord with the diagram on page 8. Define $R(x, y)$ to mean that the vertices x, y are adjacent. $R(1, 3)$ is false, but $R(1, 4)$ is true. An operation can be considered as a relation connecting three elements. For example, the operation of multiplication can be considered completely in terms of the relation $R(a, b, c)$ which means $a \cdot b = c$. The relations of a mathematical system are defined by their special properties. These may be clumsy to write down. For example, the special properties of multiplication in a group are given by the postulates on page 3. Using the notation above we see that the third postulate gives the property $R(e, a, a)$ is true. A mathematical system, then, consists of elements and relations defined among these elements.

Two mathematical systems S and T are said to be *isomorphic* if there is a 1-1 correspondence between the elements and relations of S and T such that truth of a relation in one system implies truth of the corresponding relation in the other system, and falsity of a relation in one system implies falsity of the corresponding relation in the other.

EXAMPLES. For both S and T take the set of real numbers. Let $R(x, y)$ be the relation $x < y$ for S. For the corresponding relation $R'(x', y')$ in T take $x' > y'$. S can be mapped isomorphically on T by the transformation $x' = -x$.

EXERCISE 4. Let S be the set of vertices of the cube with the relation $R(x, y)$, x and y have an edge in common. Show that this is isomorphic to the set T of faces of the octahedron where the relation $R'(x', y')$ means that the faces x', y' are adjacent. (See Figure 4.1.) Label the faces of the octahedron accordingly.

For the projective plane take the relation $R(P, \ell)$ to mean that the point P is on the line ℓ. To set up an isomorphism between two planes, use central projection from a point outside both. $R(P, \ell) \Leftrightarrow R(P', \ell')$. For ordinary Euclidean planes only parallel projection will give an isomorphism.

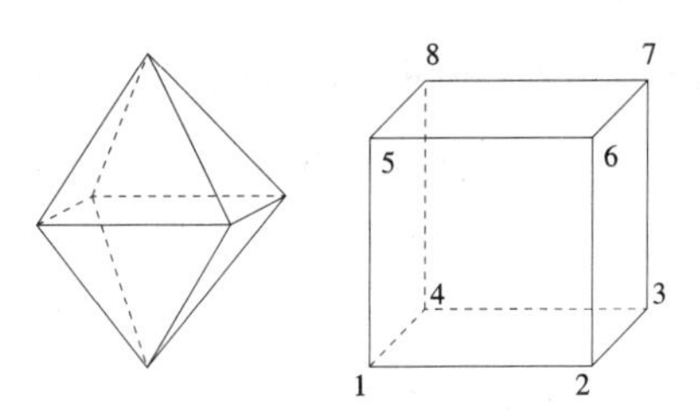

FIGURE 4.1

An *automorphism* is an isomorphism of S with itself. Hence an automorphism is a 1-1 mapping of a system onto itself which preserves the validity of the relations among its elements. Every system has at least one automorphism—the identity—the mapping which takes each element into itself.

EXAMPLES. Consider the automorphisms of the cube which preserve the validity of the relation $R(x, y)$: x, y are adjacent. The 90° rotation which takes the vertices 1,2,3,4,5,6,7,8 into the vertices 2,6,7,3,1,5,8,4 is just such an automorphism. We denote it by

$$\begin{matrix} 1 & 2 & 3 & 4 & 5 & 6 & 7 & 8 \\ 2 & 6 & 7 & 3 & 1 & 5 & 8 & 4 \end{matrix}$$

or simply by (26731584).[2] These automorphisms are called the *symmetries* of the cube and, as the nomenclature suggests, they are strongly connected with its regular geometric properties.

EXERCISE 5. Determine the 48 automorphisms of the cube. In the set of integers let $R(x, y)$ mean $x < y$. The translations

$$a \to a + n$$

are all possible automorphisms of this system. In the same set, we define $R(a, b, c)$ to mean a is between b and c.

EXERCISE 6. Find all the automorphisms of the integers which preserve the validity of the relation $R(a, b, c)$.

In addition to these examples, we introduce two important illustrations of numerical systems:

(1) The only automorphisms of the integers which preserve addition are identity and the mapping $x \to -x$.

PROOF: Put $0' = a$. Then $0 + 0 = 0 \Rightarrow a + a = a$, and consequently, $a = 0$. Furthermore, either $1' = 1$ or $1' = -1$. For put $1' = b$. It follows that

$$2' = (1 + 1)' = b + b = 2b, \quad 3' = (1 + 2)' = b + 2b = 3b, \quad \text{etc.}$$

Hence every integer is a multiple of b. Therefore b can only be one of the units, either $+1$ or -1. □

---

[2]Compare the notation on page 8.

(2) The field of real numbers possesses no automorphisms other than identity.

PROOF: The elements 0 and 1 remain fixed in the automorphisms of any field. For clearly $0' = 0$ and therefore $1' = a \neq 0$. Consequently,

$$1 \cdot 1 = 1 \Rightarrow a \cdot a = a, \quad a \neq 0 \Rightarrow a = 1.$$

It follows that the integers remain fixed. For

$$2 = 1 + 1 \to 1 + 1 = 2, \quad 3 = 2 + 1 \to 2 + 1 = 3, \quad \text{etc.},$$

and furthermore

$$\begin{matrix} n + (-n) = 0 \\ n' = n, 0' = 0 \end{matrix} \Rightarrow (-n)' = -n.$$

As an immediate consequence the rational numbers must also remain fixed: The number $x = m/n$ satisfies the relation $nx = m$. Put $x' = y$

$$\begin{matrix} n' = n \\ m' = m \end{matrix} \Rightarrow ny = m.$$

The solution of this equation is unique. Consequently, x maps onto itself. □

The ordering of the real numbers is not changed by automorphism, i.e.,

$$a < b \Rightarrow a' < b'.$$

In deriving this result we may not use limiting relations since we are concerned only with the field relations, addition and multiplication, and it is only these that need be preserved. We assert that limiting relations in a field of numbers are not always preserved by automorphism. In proof of this assertion we offer the following:

EXAMPLE. Let F be the field of numbers $a + b\sqrt{2}$ where a and b are rational. We use the result of

EXERCISE 7. Show that the mapping $a + b\sqrt{2} \to a - b\sqrt{2}$ is an automorphism of F (i.e., show that the mapping preserves addition and multiplication properties). This automorphism leaves the rational elements fixed. We can therefore approximate $\sqrt{2}$ as closely as we please by fixed elements 1, 1.4, 1.41, 1.414, ... Even so, $\sqrt{2}$ is not fixed. We see that continuity properties are not preserved in this automorphism of the field. Nevertheless, it is not difficult to prove the contrary for the field of real numbers.

It is sufficient to show that $a > 0 \Rightarrow a' > 0$. To this end we use: $a > 0 \Leftrightarrow \exists b$ such that $b^2 = a$. Hence

$$a > 0 \Rightarrow a = b \cdot b \Rightarrow a' = b' \cdot b' \Rightarrow a' > 0.$$

Consequently,

$$c > d \Rightarrow c + (-d) > 0 \Rightarrow c' + (-1)d' > 0 \Rightarrow c' > d'.$$

Thus, in any automorphism of the field of real numbers the order of the elements is preserved. Any real number is uniquely defined by its inequalities with respect to the rational numbers (Dedekind cut). Since the rational numbers remain fixed, it is clear that each element can go only into itself. The only automorphism is identity.

The field of complex numbers, on the other hand, has at least one nonidentical automorphism, $a + bi \leftrightarrow a - bi$. In fact, the set of automorphisms of the complex number field has the cardinal number

$$2^{2^{\aleph_0}}.$$

Let $p(x)$ be an irreducible polynomial over a field F, and E be an extension field of F which contains a root α of $p(x)$. We denote the smallest extension field between E and F by $F(\alpha)$, $E \subset F(\alpha) \subset F$. It has been demonstrated (p. 41) that $F(\alpha)$ is isomorphic to the field of residue classes of polynomials mod $p(x)$ under the mapping $\phi(\alpha) \leftrightarrow \phi(x)$. If $\partial[p(x)] = n + 1$, it is unnecessary to consider polynomials of degree greater than n (by the proof on p. 38). Thus any element of $F(\alpha)$ can be written in the form

$$c_0 + c_1\alpha + \cdots + c_n\alpha^n$$

where $c_0, c_1, \ldots, c_n \in F$. The sum of two such elements is at once of the same form. The product can be handled by the method of the following

EXAMPLE. The polynomial $p(x) = x^5 - x - 1$ is irreducible over the field R of rational numbers. Hence if α is a root of $p(x)$, all elements of $R(\alpha)$ may be written as above:

$$c_0 + c_1\alpha + \cdots + c_4\alpha^4.$$

The product of two such elements is a polynomial in α of degree ≤ 8. It can be reduced to the prescribed form by means of the rules

$$\alpha^5 = 1 + \alpha, \quad \alpha^6 = \alpha + \alpha^2, \quad \alpha^7 = \alpha^2 + \alpha^3, \quad \alpha^8 = \alpha^3 + \alpha^4.$$

This method is applicable to any irreducible polynomial.

The only difficulty in the demonstration of the field properties of $F(\alpha)$ lies in writing the quotient $\phi(\alpha)/\psi(\alpha)$, $\psi(\alpha) \neq 0$, as a polynomial of the prescribed form. Put

$$\phi(\alpha)/\psi(\alpha) = c_0 + c_1\alpha + \cdots + c_n\alpha^n,$$

where the coefficients $c_0, c_1, \ldots, c_n$ are to be determined. We write

$$\phi(\alpha) = \psi(\alpha)(c_0 + c_1\alpha + \cdots + c_n\alpha^n).$$

If the right side is reduced according to the rules for multiplication of two elements, we obtain

$$\phi(\alpha) = L_0 + L_1\alpha + \cdots + L_n\alpha^n,$$

where $L_0, L_1, \ldots, L_n$ are linear combinations of the c_i. Equating corresponding coefficients we obtain n linear equations in n unknowns. This system of linear equations always has a solution since the system of homogeneous equations given by $\phi(\alpha) = 0$ has only the trivial solution

$$c_0 = c_1 = \cdots = c_n = 0.$$

We have obtained a method of handling operations on the elements of $F(\alpha)$. This treatment is based on the assumption that α is the root of a given irreducible polynomial. As yet we have hardly any criteria for determining whether a given polynomial is irreducible. We will formulate such criteria later.

EXERCISE 8. Show that $x^5 - x - 1$ is irreducible over the field of rational numbers. This result can be derived from

EXERCISE 9. Prove that a polynomial with integer coefficients possesses a factorization into polynomials with integer coefficients provided it can be factored at all.

We have considered several specific examples of isomorphism between fields. Let us now analyze this situation in complete abstract generality. Assume two fields, F and $\overline{F}$, to be given together with an isomorphism which assigns to any $a \in F$ an $\bar{a} \in \overline{F}$. Thus

$$a + b = c \Leftrightarrow \bar{a} + \bar{b} = \bar{c}, \quad a \cdot b = c \Leftrightarrow \bar{a} \cdot \bar{b} = \bar{c}.$$

Therefore

$$0 + 0 = 0 \Rightarrow \bar{0} + \bar{0} = \bar{0}.$$

Hence $\bar{0}$ is the zero element of $\overline{F}$. In the same way we see that $\bar{1}$ is the unit element of $\overline{F}$. From these results we shall see that subtraction and division properties are also preserved. We have

$$a + (-a) = 0, \quad \bar{a} + \overline{(-a)} = \bar{0},$$

whence

$$-\bar{a} = \overline{(-a)}.$$

Consequently,

$$\overline{\left(\frac{a}{b}\right)} = \overline{ab^{-1}} = \bar{a}(\bar{b})^{-1} = \frac{\bar{a}}{\bar{b}}.$$

The isomorphism between F and $\overline{F}$ can be extended in a very natural way to the rings $F(x)$ and $\overline{F}(x)$ of polynomials, over the respective fields. Given the polynomial

$$f(x) = a_0 + a_1 x + \cdots + a_n x^n,$$

define

$$\bar{f}(x) = \bar{a}_0 + \bar{a}_1 x + \cdots + \bar{a}_n x^n.$$

Thus we have provided a 1-1 correspondence between the elements of $F(x)$ and $\overline{F}(x)$. It is an easy matter to show that the isomorphism between F and $\overline{F}$ extends to $F(x)$ and $\overline{F}(x)$, i.e., to show that

$$f(x) + g(x) \leftrightarrow \bar{f}(x) + \bar{g}(x),$$
$$f(x) \cdot g(x) \leftrightarrow \bar{f}(x) \cdot \bar{g}(x).$$

The usual properties of polynomials all go over in this way; e.g., a polynomial has the same degree as its image. The irreducibility of polynomials is preserved. For suppose $p(x)$ is irreducible and $\bar{p}(x)$ is not. Then we would have

$$\bar{p}(x) = \bar{a}(x) \cdot \bar{b}(x) \Rightarrow p(x) = a(x) \cdot b(x),$$

which contradicts the hypothesis. We have shown that F and $\overline{F}$ behave in exactly the same way. The difference between F and $\overline{F}$ is like a difference of color—a very unessential distinction for fields.

THEOREM 4.6 *Let $p(x)$ be an irreducible polynomial over F, and $\bar{p}(x)$ the corresponding polynomial in the isomorphic field $\overline{F}$. Let α and $\bar{\alpha}$, respectively, be roots (obtained by any means whatsoever) of these polynomials. The isomorphism between F and $\overline{F}$ can then be "extended" to the fields $F(\alpha)$ and $\overline{F}(\bar{\alpha})$. (The mapping of $F(\alpha)$ on $\overline{F}(\bar{\alpha})$ is called an* extension *of the mapping of the ground fields if it contains the given correspondence between the elements of F and $\overline{F}$.)*

PROOF: The elements of $F(\alpha)$ are polynomials

$$\theta = c_0 + c_1\alpha + \cdots + c_{n-1}\alpha^{n-1},$$

where $n = \partial[p(x)]$. Map these elements onto the corresponding elements of $\overline{F}(\bar{\alpha})$,

$$\bar{\theta} = \bar{c}_0 + \bar{c}_1\bar{\alpha} + \cdots + \bar{c}_{n-1}\bar{\alpha}^{n-1}.$$

For addition, we have

$$\overline{\theta_1 + \theta_2} = \bar{\theta}_1 + \bar{\theta}_2.$$

For multiplication, however, we may have to reduce the degree. Put $\theta_1 = \phi_1(\alpha)$, $\theta_2 = \phi_2(\alpha)$ where $\partial[\phi_1(x)], \partial[\phi_2(x)] < n$:

$$\phi_1(x) \cdot \phi_2(x) = q(x)p(x) + r(x), \quad \partial[r(x)] < n,$$

whence

$$\phi_1(\alpha) \cdot \phi_2(\alpha) = r(\alpha).$$

We must prove

$$\bar{\phi}_1(\bar{\alpha}) \cdot \bar{\phi}_2(\bar{\alpha}) = \bar{r}(\bar{x}).$$

Since $F(x)$ and $\overline{F}(x)$ are isomorphic, we have

$$\overline{\phi_1(x) \cdot \phi_2(x)} = \overline{q(x)p(x) + r(x)} = \bar{\phi}_1(x) \cdot \bar{\phi}_2(x) = \bar{q}(x)\bar{p}(x) + \bar{r}(x).$$

Putting $x = \bar{\alpha}$ we obtain

$$\bar{\phi}_1(\bar{\alpha}) \cdot \bar{\phi}_2(\bar{\alpha}) = \bar{r}(\bar{\alpha}),$$

and the proof is complete. □

EXAMPLES. Let us consider isomorphisms of the field R of rational numbers with itself. The only automorphism of R is identity. It follows that any extension of this mapping by means of a root of the irreducible polynomial $p(x)$ leaves R fixed.

(a) Take $p(x) = x^2 - 2$. This polynomial has the roots $\sqrt{2}$ and $-\sqrt{2}$ in the field of real numbers. $R(\sqrt{2})$ consists of elements of the form $a + b\sqrt{2}$, $R(-\sqrt{2})$ of elements of the form $a - b\sqrt{2}$, $a, b \in R$. Clearly both extensions give the same field. By Theorem 4.6 this proves an automorphism of $R(\sqrt{2})$. We have demonstrated the result of Exercise 7. However, the method does not generally give an automorphism. Consider the example:

(b) $p(x) = x^3 - 2$. Let $a = \sqrt[3]{2}$ be the real cube root of 2, and $\bar{\alpha}$ one of the complex roots. $R(\sqrt[3]{2})$ consists only of real elements but $R(\bar{\alpha})$ contains elements which are complex. The fields $R(\bar{\alpha})$ and $R(\sqrt[3]{2})$ are isomorphic but clearly not the same.

(c) The points of the complex plane corresponding to the nth roots of unity are the vertices of a regular n-sided polygon. This fact allows us to handle polygons in a very convenient manner.

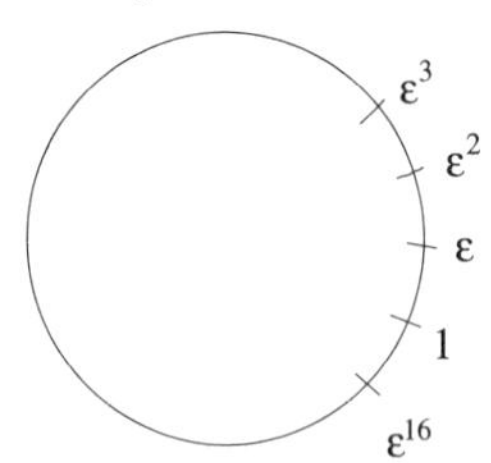

Consider, for example, the regular polygon of 17 sides. Its vertices are given by the roots of the polynomial $x^{17} - 1$, which is reducible in R (1 is clearly a root). Factoring out $x - 1$ we obtain the polynomial $p(x) = 1 + x + x^2 + \cdots + x^{15} + x^{16}$. Assume for the present, without proof, that this polynomial is irreducible. Let ε be the complex number corresponding to the first vertex counterclockwise from 1. The successive vertices are given by $\varepsilon, \varepsilon^2, \dots, \varepsilon^{16}, \varepsilon^{17} = 1$. The first 16 powers of ε are obviously the roots of $p(x)$. Now consider the extension fields $R(\varepsilon)$ and $R(\varepsilon^3)$. $R(\varepsilon^3)$ contains ε since $(\varepsilon^3)^6 = \varepsilon^{18} = \varepsilon$. Hence $R(\varepsilon^3) \supset R(\varepsilon)$. But $R(\varepsilon) \supset R(\varepsilon^3)$ and therefore $R(\varepsilon) = R(\varepsilon^3)$. The isomorphism is actually an automorphism which maps $\theta \in R(\varepsilon)$ onto $\bar{\theta} \in R(\varepsilon)$ where if

$$\theta = c_0 + c_1\varepsilon + c_2\varepsilon^2 + \cdots + c_{15}\varepsilon^{15}$$

then

$$\bar{\theta} = c_0 + c_1\varepsilon^3 + c_2\varepsilon^6 + \cdots + c_{15}\varepsilon^{11},$$

where each power of ε above the 15th is reduced by using $\varepsilon^{17} = 1$ and $p(\varepsilon) = 0$; for example,

$$(\varepsilon^3)^{11} = (\varepsilon)^{33} = (\varepsilon)^{16} = -(1 + \varepsilon + \cdots + \varepsilon^{15}).$$

If, instead of ε^3, we take ε^ν where $\nu \not\equiv 0 \pmod{17}$ we can find an x such that $\nu x \equiv 1 \pmod{17}$. Hence $(\varepsilon^\nu)^x = \varepsilon$. Therefore, by the same reasoning as above, each value of ν gives a different automorphism of the field $R(\varepsilon)$. It will be shown that the nature of these 16 automorphisms permits us to see that the polygon of 17 sides possesses a construction in the Euclidean sense. By the same methods we will be able to see that no construction exists for the 13-sided polygon.

CHAPTER 5

Galois Theory

5.1. Splitting Fields

Let $f(x)$ be any polynomial over a field F.

THEOREM 5.1 *There is an extension field $E \supset F$ such that $f(x)$ is the product of linear factors in E. It is then said that $f(x)$* splits *in the field E.*

PROOF: The polynomial $f(x)$ possesses a unique factorization into irreducible factors in F. Thus we may write

$$f(x) = c(x - \alpha_1)(x - \alpha_2) \cdots (x - \alpha_r)p_1(x)p_2(x) \cdots p_s(x)$$

where $\alpha_1, \alpha_2, \ldots, \alpha_r$ are the roots of $f(x)$ in F and $p_1(x), p_2(x), \ldots, p_s(x)$ are the irreducible factors of degree higher than 1.

If $s = 0$ then $f(x)$ splits in F and we need go no further. Otherwise solve $p_1(x) = 0$ in any extension field. Let α be a root of $p_1(x)$. In the field $F(\alpha)$, $p_1(x)$ has a linear factor,

$$p_1(x) = (x - \alpha)q(x).$$

Now take $F(\alpha)$ as the ground field and factor $f(x)$ in $F(\alpha)$. The new factorization possesses at least one additional linear factor, namely $x - \alpha$, and perhaps more. If $f(x)$ splits in $F(\alpha)$, $F(\alpha)$ is the desired extension field. If not we may repeat the argument and obtain an extension field of $F(\alpha)$ in which $f(x)$ has at least one additional factor. Clearly, the process terminates, and we arrive in a finite number of steps at a field E in which $f(x)$ splits into linear factors. □

An extension field $E \supset F$ which is obtained by this method is called a *splitting field* of $f(x)$ over F.

THEOREM 5.2 *Let $f(x)$ be any polynomial in F and Ω any extension field $\Omega \supset F$, in which $f(x)$ can be split into linear factors,*

$$f(x) = c(x - \alpha_1)(x - \alpha_2) \cdots (x - \alpha_n).$$

The smallest field in Ω for which $f(x)$ splits is the field E obtained by the method of Theorem 5.1.

PROOF: If there is a field E, $\Omega \supset E \supset F$, in which $f(x)$ splits, then E must contain the elements $\alpha_1, \alpha_2, \ldots, \alpha_n$. Since E contains F, it contains all possible combinations of sums and products of the α_i with the elements of F; i.e., E contains all polynomials in the α_i.[1] If this set of polynomials is a field—and we shall prove that it is—it is certainly the smallest splitting field of $f(x)$ between Ω and F.

Consider the set of all polynomials

$$\phi(\alpha_1, \alpha_2, \ldots, \alpha_n)$$

with coefficients in F. We now prove that this is the field E of the previous theorem. Since α_1 is algebraic over F the polynomials $\phi(\alpha_1)$ over F form a field $F(\alpha_1)$. Furthermore, since $F \subset F(\alpha_1)$, α_2 is algebraic in $F(\alpha_1)$. (It satisfies the equation $f(x) = 0$ over $F(\alpha_1)$.) Therefore, the set of all polynomials in α_2 whose coefficients are elements of $F(\alpha_1)$—that is, polynomials in α_1—form a field $F(\alpha_1, \alpha_2)$. It follows by induction that

$$E = F(\alpha_1, \alpha_2, \ldots, \alpha_n),$$

the set of all polynomials in the α_i. Note that this field is, in fact, the same as the field obtained in Theorem 5.1. □

The field E is called *the* splitting field of $f(x)$ between F and Ω.

EXAMPLE. Let F be the field R of rational numbers, Ω the field of real numbers. Take

$$f(x) = (x^2 - 2)(x^2 - 3) = (x + \sqrt{2})(x - \sqrt{2})(x + \sqrt{3})(x - \sqrt{3}).$$

Clearly, $E = R(\sqrt{2}, \sqrt{3})$ and hence consists of elements

$$(a + b\sqrt{2}) + (c + d\sqrt{2})\sqrt{3} = a + b\sqrt{2} + c\sqrt{3} + d\sqrt{6},$$

$a, b, c, d \in R$.

The dimension of the vector space E of the polynomials[2] in the α_i over F is called the degree of E over F. Thus the degree of $R(\sqrt{2}, \sqrt{3})$ is at most 4.

EXERCISE 1. Show that the degree of $R(\sqrt{2}, \sqrt{3})$ over R is exactly 4.

THEOREM *Let $f(x)$ be a polynomial over F. Any two splitting fields of $f(x)$ over F are isomorphic.*

We shall prove this result in the more general form:

THEOREM 5.3 *If $f(x)$ is any polynomial over F and $\bar{f}(x)$ is the corresponding polynomial over an isomorphic field $\overline{F}$, and if E is the splitting field of $f(x)$, $\overline{E}$ of $\bar{f}(x)$, then the isomorphism between F and $\overline{F}$ can be extended to E and $\overline{E}$.*

PROOF: Write the factorization of $f(x)$ into irreducible factors over F:

$$f(x) = c(x - \alpha_1)(x - \alpha_2) \cdots (x - \alpha_r)p_1(x)p_2(x) \cdots p_s(x)$$

[1] A polynomial in two variables x, y is a polynomial in y whose coefficients are polynomials in x. A polynomial in the $n+1$ variables $x_1, x_2, \ldots, x_{n+1}$ is a polynomial in x_{n+1} whose coefficients are polynomials in the n variables $x_1, x_2, \ldots, x_n$.

[2] Cf. example on p. 16.

where the $p_i(x)$ are the irreducible factors of degree higher than 1. Since F and $\overline{F}$ are isomorphic this gives the factorization

$$\bar{f}(x) = \bar{c}(x - \bar{\alpha}_1)(x - \bar{\alpha}_2) \cdots (x - \bar{\alpha}_r)\bar{p}_1(x)\bar{p}_2(x) \cdots \bar{p}_s(x)$$

of $\bar{f}(x)$ into irreducible polynomials over $\overline{F}$. Let n be the degree of $f(x)$ and r the number of linear terms in the factorization. In the proof of the theorem we use an induction in the following form:

If a theorem is true for n and if the truth of the theorem for $r+1$ implies its truth for r, then it is true for all $r \leq n$.

If $r = n$, the polynomial $f(x)$ splits into linear factors in F. Moreover, $\bar{f}(x)$ splits in $\overline{F}$ in exactly the same way. Consequently, $E = F$ and $\overline{E} = \overline{F}$. We have established the first step in the induction.

Assume that the theorem has been proved for polynomials having at least $r+1$ linear factors, $r < n$. Suppose now that $f(x)$ has r linear factors in F. Since $p_1(x)$ splits in E, $\bar{p}_1(x)$ in $\overline{E}$, they have roots $\alpha_{r+1} \in E$, $\overline{\alpha_{r+1}} \in \overline{E}$, respectively. Construct the extension fields $F(\alpha_{r+1})$ and $\overline{F}(\overline{\alpha_{r+1}})$ and extend the isomorphism of F and $\overline{F}$ to these fields by means of the transformation $\alpha_{r+1} \leftrightarrow \overline{\alpha_{r+1}}$ (Theorem 4.6, Chapter 4). Since the isomorphism of $F(\alpha_{r+1})$ and $\overline{F}(\overline{\alpha_{r+1}})$ contains that of F and $\overline{F}$, the mapping $f(x) \leftrightarrow \bar{f}(x)$ is retained. $F(\alpha_{r+1})$ and $\overline{F}(\overline{\alpha_{r+1}})$ are now taken as the ground fields. We again factor $f(x)$ and $\bar{f}(x)$ but now we obtain at least one additional linear factor $(x - \alpha_{r+1})$ in $F(\alpha_{r+1})$ and $(x - \overline{\alpha_{r+1}})$ in $\overline{F}(\overline{\alpha_{r+1}})$. Thus $f(x)$ possesses at least $r+1$ factors in $F(\alpha_{r+1})$. Furthermore, it is clear that E is the splitting field of $f(x)$ over $F(\alpha_{r+1})$ since $f(x)$ splits in E and certainly does not split in any smaller field between E and F. The same results apply to $\bar{f}(x)$. Hence, by the induction hypothesis, the isomorphism between $F(\alpha_{r+1})$ and $\overline{F}(\overline{\alpha_{r+1}})$ can be extended to E and $\overline{E}$. Since E and $\overline{E}$ are the respective splitting fields of $f(x)$ over F and $\bar{f}(x)$ over $\overline{F}$, the induction is complete. □

The reason for proving the more general theorem is that in order to be able to use induction one must make the statement of the theorem for $r+1$ identical with that for n.

5.2. Automorphisms of the Splitting Field

Let F be a field, E the splitting field of $f(x)$ over F. What are all the automorphisms of E that leave F fixed? A partial answer is provided at once by

LEMMA 5.4 *If $f(x)$ possesses a nonlinear irreducible factor $p(x)$ with distinct roots $\alpha_1, \alpha_2, \ldots, \alpha_n \in E$, then by means of the transformation $\alpha_i \leftrightarrow a_j$, $i \neq j$, we obtain an isomorphism of $F(\alpha_i)$ with $F(\alpha_j)$. By Theorem 5.3 this isomorphism can then be extended to give a nontrivial (i.e., nonidentical) automorphism of E. Since F remains fixed in the automorphism of $F(\alpha_i)$ with $F(\alpha_j)$, it remains fixed in the automorphism of E.*

EXAMPLE. Take $f(x) = (x^2-2)(x^2-3)$ over the field R of rational numbers. A splitting of $f(x)$ can be obtained in the field of real numbers, for we may write

$$p_1(x) = x^2 - 2 = (x + \sqrt{2})(x - \sqrt{2}),$$
$$p_2(x) = x^2 - 3 = (x + \sqrt{3})(x - \sqrt{3}).$$

By means of the roots $\pm\sqrt{2}$ of $p_1(x)$ we can extend R in two ways to obtain the field $R(\sqrt{2})$. The automorphisms of $R(\sqrt{2})$ can then be extended to the splitting field E of $f(x)$. Now consider $R(\sqrt{2})$ as the ground field. $R(\sqrt{2})$ cannot be the splitting field of $p_2(x) = x^2 - 3$ by the result of

EXERCISE 2. Demonstrate the impossibility of finding $a, b \in R$ such that

$$a + b\sqrt{2} = \sqrt{3}.$$

If $R(\sqrt{2})$ is extended by either of the roots $\pm\sqrt{3}$ of $p_2(x)$, we obtain the splitting field $E = R(\sqrt{2}, \sqrt{3})$. Thus we have two automorphisms of E which leave $R(\sqrt{2})$ fixed. Combining our results we obtain four[3] automorphisms of E,

$\sqrt{2} \to$	$\sqrt{2}$	$-\sqrt{2}$	$\sqrt{2}$	$-\sqrt{2}$
$\sqrt{3} \to$	$\sqrt{3}$	$\sqrt{3}$	$\sqrt{3}$	$-\sqrt{3}$

which leave R fixed.

If Ω is an extension field of F, it is conceivable that the set of all automorphisms which leave F fixed leave other elements fixed than those of F.

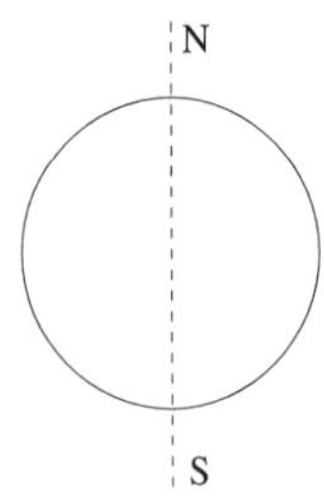

EXAMPLE. Consider the automorphisms of the sphere which preserve the distances of points. All automorphisms which leave two diametrically opposite points fixed also leave fixed all points on their common diameter. We shall show that no other elements than those of F remain fixed if Ω is the splitting field of a certain very general kind of polynomial over F.

THEOREM 5.5 *Factor $f(x)$ into irreducible polynomials over F:*

[1] $$f(x) = c(x - \alpha_1)(x - \alpha_2) \cdots (x - \alpha_r) P_1(x) P_2(x) \cdots P_s(x),$$

where the $P_i(x)$ are factors of degree higher than 1. If

(a) *E is the splitting field of $f(x)$ over F, and*

(b) *no linear factor appears more than once in the splitting of $P_i(x)$ in E, $i = 1, 2, \ldots, n$ (i.e., the polynomial $P_i(x)$ has no multiple roots in E),*

then no elements but those of F remain fixed under all the automorphisms of E which leave F fixed.

PROOF: We use an induction in the manner of Theorem 5.3. Put $n = \partial[f(x)]$ and let r be the number of linear terms in the factorization [1]. If $r = n$ then $E = F$ and the theorem is trivially true. Assume that the theorem is true in all cases where there are at least $r + 1$ linear factors:

[3] Cf. the result of Exercise 1 on p. 50.

If α_{r+1} is a root of $P_1(x)$ the field F can be extended to the field $F(\alpha_{r+1})$ in which $f(x)$ has $r+1$ linear factors at least. To derive the factorization of $f(x)$ in $F(\alpha_{r+1})$, first obtain the factorization [1] of $f(x)$ in F and then factor the $P_i(x)$ into irreducible polynomials over $F(\alpha_{r+1})$. This method yields

$$f(x) = c(x-\alpha_1)\cdots(x-\alpha_r)(x-\alpha_{r+1})(x-\beta_1)\cdots(x-\beta_\mu)Q_1(x)\cdots Q_\nu(x)$$

where each of the $Q_k(x)$ is a nonlinear irreducible polynomial in the factorization of one of the $P_i(x)$. Now, as we observed in the proof of Theorem 5.3:

(1) E is the splitting field of $f(x)$ over $F(\alpha_{r+1})$.

Furthermore, since $Q_k(x)|P_i(x)$

(2) The splitting of $Q_k(x)$ in E contains no repeated factors.

Otherwise a factor would appear more than once in the splitting of $Q_k(x)$ and hence in the splitting of $P_i(x)$, contrary to (b). We have demonstrated that the polynomial $f(x)$ over $F(\alpha_{r+1})$ satisfies the conditions of the theorem. Thus, according to the induction hypothesis, if $\theta \in E$ is left fixed by all the automorphisms of E which leave $F(\alpha_{r+1})$ fixed, then $\theta \in F(\alpha_{r+1})$.

Now suppose that θ remains fixed under all the automorphisms of E that leave F fixed. It follows that θ remains fixed under all the automorphisms that leave $F(\alpha_{r+1}) \in F$ fixed. Consequently, $\theta \in F(\alpha_{r+1})$. Put $\partial[P_1(x)] = t$. It follows that

$$\theta = c_0 + c_1\alpha_{r+1} + \cdots + c_{t-1}\alpha_{r+1}^{t-1}$$

for some $c_0, c_1, \ldots, c_{t-1} \in F$. According to (b) the factorization of $P_1(x)$ in E contains no repeated factors. Therefore we may write

$$P_1(x) = (x-\alpha_{r+1})(x-\alpha_{r+2})\cdots(x-\alpha_{r+t}),$$

where $\alpha_{r+1}, \alpha_{r+2}, \ldots, \alpha_{r+t}$ are the t distinct roots of $P_1(x)$ in E. It follows by Lemma 5.4 that the t transformations $\alpha_{r+1} \leftrightarrow \alpha_{r+j}$ provide automorphisms of E which leave F fixed. But, since θ is fixed under all such automorphisms it may be written in t different ways:

$$\begin{aligned}\theta = c_0 + c_1\alpha_{r+1} + \cdots + c_{t-1}\alpha_{r+1}^{t-1} &= c_0 + c_1\alpha_{r+2} + \cdots + c_{t-1}\alpha_{r+2}^{t-1}\\ &\;\vdots\\ &= c_0 + c_1\alpha_{r+t} + \cdots + c_{t-1}\alpha_{r+t}^{t-1}.\end{aligned}$$

Thus the polynomial

$$\phi(x) = (c_0 - \theta) + c_1 x + \cdots + c_{t-1}x^{t-1}$$

of degree at most $t-1$ possesses t distinct roots in E. It follows that $\phi(x) = 0$ whence all the coefficients are zero. From $\theta - c_0 = 0$ we see that $\theta \in F$. □

5.3. The Characteristic of a Field

The elements of a field F form an additive group. Consider the cyclic subgroup formed by repeated additions and subtractions of the unit element:

[*] $\qquad \ldots, -1-1, -1, 0, 1, 1+1, 1+1+1, \ldots.$

We denote these elements by

$$\dots, -2, -1, 0, 1, 2, 3, \dots,$$ [4]

respectively. The symbol 17, say, now has two possible meanings, the integer 17 or the field element

$$1 + 1 + \cdots + 1 \quad (17 \text{ times}).$$

This ambiguity of meaning will not lead to confusion since the ordinary rules for addition and multiplication of integers apply to the corresponding field elements. Thus

$$2 + 3 = 5; (1 + 1) + (1 + 1 + 1) = 1 + 1 + 1 + 1 + 1;$$
$$2 \cdot 3 = 6; (1 + 1) \cdot (1 + 1 + 1) = (1 + 1 + 1) + (1 + 1 + 1).$$

We distinguish two cases:

Case 1. The elements in the sequence [$*$] are distinct. The field F is then said to be of *characteristic zero*. Clearly F contains a subring which is isomorphic to the integers. For the same reason it contains a subfield, isomorphic to the rational numbers, which consists of elements m/n, $n \neq 0$.

Case 2. Some element appears twice in the sequence [$*$]. In that case, there is a period, call it d. The element denoted by d is zero.

LEMMA 5.6 *If the additive cyclic group generated by the unit element of F has a period d, then d is prime.*

PROOF: Assume d is not prime. Then $d = a \cdot b$ where a and b are positive and less than d. But

$$d = 0 \Rightarrow ab = 0 \Rightarrow \text{ either } a = 0 \text{ or } b = 0$$

contradicting the hypothesis that d is least. □

If the period of the unit element is the prime p, then any nonzero $a \in F$ has the period p.

PROOF: We may write the sum

$$a + a + \cdots + a \quad (n \text{ times})$$

in the form

$$1 \cdot a + 1 \cdot a + \cdots + 1 \cdot a = (1 + 1 + \cdots + 1)a = na.$$

Thus, since $p = 0$, $p \cdot a = 0$. We have only to show that n is least. But this is clear since

$$\left.\begin{array}{r} n \cdot a = 0 \\ a \neq 0 \end{array}\right\} \Rightarrow n = 0 \Rightarrow p | n.$$

□

[4] We are at liberty to call anything by any name we please. If a piece of chalk is called Emma does that mean it is a human being? (No, it is a seagull.) Cf. Christian Morgenstern, *Die Moewen*,

Die Moewen sehen alle aus
Als ob sie Emma heissen ....

If the nonzero elements of a field F have the period p, F is said to be of *characteristic* p.

5.4. Derivative of a Polynomial: Multiple Roots

To each polynomial $f(x)$ over F we associate another polynomial $f'(x)$, the *derivative* of $f(x)$. If

$$f(x) = a_0 + a_1 x + \cdots + a_\nu x^\nu + \cdots + a_n x^n,$$

we define

$$f'(x) = a_1 + 2a_2 x + \cdots + \nu a_\nu x^{\nu-1} + \cdots + n a_n x^{n-1}$$

where the coefficient νa_ν of $x^{\nu-1}$ is the sum

$$\nu a_\nu = a_\nu + a_\nu + \cdots + a_\nu \quad (\nu \text{ times}).$$

The properties of derivatives which are familiar from analysis do not necessarily have validity here. For example, $f'(x) = 0$ does not always imply that $f(x)$ is a constant; e.g., if we set $f(x) = x^{17}$ in a field of characteristic 17 then

$$f'(x) = 17x^{16} = 0.$$

The ordinary rules for operating with derivatives, however, remain the same.

It is easy to verify that taking the derivative is a *linear operation*, i.e.,

$$[af(x) + bg(x)]' = af'(x) + bg'(x)$$

where $a, b \in F$.

LEMMA 5.7 *For the derivative of a product we have the usual rule*

$$[f(x) \cdot g(x)]' = f'(x) \cdot g(x) + f(x) \cdot g'(x).$$

PROOF: If the statement of the lemma is true for two choices f_1 and f_2 of $f(x)$, it is true for any linear combination $af_1 + bf_2$, $a, b \in F$: Assume

$$(f_1 g)' = f_1' g + f_1 g', \quad (f_2 g)' = f_2' g + f_2 g'.$$

It follows that

$$\begin{aligned}[(af_1 + bf_2)g]' = a(f_1 g)' + b(f_2 g)' &= (af_1' + bf_2')g + (af_1 + bf_2)g' \\ &= (af_1 + bf_2)'g + (af_1 + bf_2)g'.\end{aligned}$$

Since a polynomial is a linear combination of powers of x, it is sufficient to take $f(x)$ to be a power of x. Moreover, since the product of $f(x)$ with $g(x)$ would then be a linear combination of products of powers of x, we need only prove the lemma for products of two powers of x. Set

$$f(x) = x^r \quad \text{and} \quad g(x) = x^s.$$

This yields

$$(fg)' = (x^{r+s})' = (r+s)x^{r+s-1} = (rx^{r-1})x^s + x^r(sx^{s-1}).$$ □

This lemma may now be used for the proof of the following:

LEMMA 5.8 *Let α be a multiple root of $f(x)$; then it is a root of $f'(x)$. Conversely, if α is a simple root of $f(x)$ then $f'(\alpha) \neq 0$.*

PROOF: To say that α is a multiple root of $f(x)$ means that the factor $x - \alpha$ appears at least twice in the factorization of $f(x)$ into irreducible polynomials over $F(\alpha)$. Consequently, we may write

$$f(x) = (x - \alpha)^2 \phi(x)$$

in the extension field $F(\alpha)$. Applying Lemma 5.7 we find that

$$f'(x) = 2(x - \alpha)\phi(x) + (x - \alpha)^2 \phi'(x),$$

whence $f'(\alpha) = 0$. Conversely, if α is a simple root we may write

$$f(x) = (x - \alpha)\phi(x) \quad \text{where } \phi(\alpha) \neq 0.$$

Lemma 5.7 yields

$$f'(x) = \phi(x) + (x - \alpha)\phi'(x)$$

whence

$$f'(\alpha) = \phi(\alpha) \neq 0. \qquad \square$$

We now use Lemma 5.8 to determine the nature of those irreducible polynomials which may have multiple roots.

Let $P(x)$ be a polynomial irreducible over a field F. Construct $F(\alpha)$, the extension of F by a root α of $P(x)$. If the multiplicity of α is greater than 1, it follows that $P'(\alpha) = 0$ or $P'(x)$ is another equation in F for α. But we assumed that $P(x)$ is irreducible; therefore $P(x) | P'(x)$ (Lemma 4.3, p. 40). However, by definition, the degree of $P'(x)$ is less than that of $P(x)$ and necessarily we have $P'(x) = 0$. Thus if we set

$$P(x) = a_0 + a_1 x + \cdots + a_n x^n, \quad a_n \neq 0,$$

we obtain

$$P'(x) = a_1 + 2a_2 x + 3a_3 x^2 + \cdots + n a_n x^{n-1}.$$

We conclude that $P(x)$ can have a multiple root only if

$$a_1 = 2a_2 = 3a_3 = \cdots = n a_n = 0.$$

But $n \geq 2 > 0$ and $a_n \neq 0$. Thus if F is a field of characteristic zero it is impossible that $n a_n = 0$. This result yields the

COROLLARY *An irreducible polynomial over a field of characteristic zero can have only simple roots.*

Assume now that F has the characteristic p. Under what conditions will the polynomial $P'(x)$ be zero? If $a_\nu x^\nu$ is the general term of $P(x)$, we must have $\nu a_\nu = 0$. Thus either $a_\nu = 0$ or $\nu = 0$; for each nonzero coefficient a_ν, the index ν must be a multiple of p. We conclude that $P(x)$ may be rewritten in the form

$$P(x) = c_0 + c_1 x^p + c_2 x^{2p} + \cdots + c_m x^{mp}.$$

Setting

$$f(x) = c_0 + c_1 x + \cdots + c_m x^m$$

we obtain $P(x) = f(x^p)$. It may very well happen that $f(x)$ has the same form as P, i.e., $f(x) = g(x^p)$ or $P(x) = g(x^{p^2})$. However, it is clear that there is a largest integer r for which $P(x)$ can be expressed in the form $P(x) = \phi(x^{p^r})$. Evidently $\phi'(x) \neq 0$, for otherwise we would write $\phi(x) = \psi(x^p)$ and hence r would not be the largest. Furthermore, $\phi(x)$ is irreducible, for otherwise

$$\phi(x) = g(x)h(x) \Rightarrow \phi(x^{p^r}) = g(x^{p^r})h(x^{p^r}) = P(x),$$

which contradicts the assumption that $P(x)$ is irreducible.

To recapitulate: $\phi(x)$ is irreducible and $\phi'(x) \neq 0$; the roots of $\phi(x)$ are then all distinct. Take for the domain of the discussion the splitting field of $P(x)\phi(x)$, i.e., the field which contains all roots of both polynomials. Denoting the roots of $\phi(x)$ by $\beta_1, \beta_2, \dots, \beta_s$, we may write

$$\phi(x) = (x - \beta_1)(x - \beta_2) \cdots (x - \beta_s)$$

where no factor is repeated. $P(x)$ may now be written

$$P(x) = \prod_{i=1}^{s} (x^{p^r} - \beta_i).$$

It therefore suffices to discuss these factors.

Consider the equation

$$x^{p^r} - \beta_i = 0$$

for the $(p^r)^{\text{th}}$ roots of β_i. We shall demonstrate that the equation has only one solution which must therefore appear with multiplicity p^r. It is first necessary to prove the

PROPOSITION *For any elements a, b in a field of characteristic p*

$$(a \pm b)^p = a^p \pm b^p.$$

PROOF: Write the binomial expansion

$$\begin{aligned}(a \pm b)^p &= (a \pm b)(a \pm b) \cdots (a \pm b) \quad (p \text{ times}) \\ &= a^p \pm \binom{p}{1} a^{p-1} b + \cdots (\pm 1)^i \binom{p}{i} a^{p-i} b^i + \cdots + (\pm 1)^p b^p,\end{aligned}$$

$(1 \leq i \leq p)$, where $\binom{p}{i} = \frac{p!}{i!(p-1)!}$ is an integer. Since $p \nmid i!(p-i)!$ it follows that $p | \binom{p}{i}$. If p is an odd prime, the proposition follows at once. If p is even, i.e., $p = 2$, then $-1 = +1$ and the proof is complete. The proposition is obviously not true for nonprimes;[5] e.g., the coefficient 6 appears in the expansion of $(a+b)^4$. $\square$

We can easily generalize the proposition to include the statement

$$(a \pm b)^{p^n} = a^{p^n} \pm b^{p^n}.$$

It has already been proved for $n = 1$. We have only to show that the truth of the statement for n implies its truth for $n + 1$. Assume

$$(a \pm b)^{p^n} = a^{p^n} \pm b^{p^n}.$$

[5]Editor's note: More precisely, for a nonprime p, it may happen that $p \nmid \binom{p}{i}$.

Then for $n+1$,

$$(a \pm b)^{p^{n+1}} = (a \pm b)^{p^n \cdot p} = [(a \pm b)^{p^n}]^p = [a^{p^n} \pm b^{p^n}]^p = a^{p^{n+1}} \pm b^{p^{n+1}}.$$

We note further that

$$[(a+b)+c]^p = (a+b)^p + c^p = a^p + b^p + c^p.$$

REMARK. It follows from this last result that if m is an integer in a field of characteristic p, i.e.,

$$m = 1 + 1 + \cdots + 1 \quad (m \text{ times}),$$

then

$$m^p = 1^p + 1^p + \cdots + 1^p = m.$$

But this is a simple generalization of the famous Fermat theorem in arithmetic:

$$m^p \equiv m \pmod{p}.$$

Now let us return to the problem: The splitting field of $P(x)\phi(x)$ is also of characteristic p since $1 \in F$ is an element of any extension field. It follows that the extraction of the $(p^r)^{\text{th}}$ roots of β_i gives a unique result. For if α_i is a root of $x^{p^r} - \beta_i$ then $\alpha_i^{p^r} = \beta_i$. Consequently,

$$x^{p^r} - \beta_i = x^{p^r} - \alpha_i^{p^r} = (x - \alpha_i)^{p^r}.$$

The splitting of $P(x)$ may now be written in the form

$$P(x) = \left[\prod_{i=1}^{s}(x - \alpha_i)\right]^{p^r}$$

which displays the fact that all roots appear with equal multiplicity p^r.

EXERCISE 3. Verify that the multiplicity p^r of the roots of an irreducible polynomial $P(x)$ over a field of characteristic p is the exponent of the greatest common divisor of the nonconstant terms of $P(x)$.

5.5. The Degree of an Extension Field

Assume given a ground field F and let E be any extension field of F. For the present disregard the general multiplication in E and utilize only multiplication by elements of F. From this point of view E is a vector space over F; namely, E is an additive commutative group, closed with respect to multiplication by elements of F, and if $a, b \in F$, $A, B \in E$, then the postulates of a vector space (p. 15) are automatically satisfied. The dimension of this vector space is called the *degree* of E over F and is denoted by (E/F). If the dimension is infinite, the degree is said to be infinite. If E is spanned by some finite number n of its elements, then $(E/F) \leq n$ (see the corollary on p. 17). Moreover, if these elements are linearly independent, the degree is precisely n. To say that E is spanned by a certain number n of its elements,

$$\omega_1, \omega_2, \ldots, \omega_n,$$

means that any element A in E can be expressed as a linear combination

$$A = a_1\omega_1 + a_2\omega_2 + \cdots + a_n\omega_n$$

where $a_1, a_2, \ldots, a_n \in F$. We then say that the elements $\omega_1, \omega_2, \ldots, \omega_n$ *generate* the field E. If, further, the ω_i are linearly independent, they are said to form a *linear basis* or simply a basis of E over F. In that case, $(E/F) = n$.

EXAMPLE. Let F be the field of real numbers, E the field of complex numbers. E consists of numbers of the form $a + bi$ where a, b are real. Thus i and 1 are generators of E. But in addition,

$$a + bi = 0 \Rightarrow a = b = 0;$$

i and 1 are linearly independent. We conclude thereby that $(E/F) = 2$.

This example is a special case of

LEMMA 5.9 *Let $P(x)$ be any irreducible polynomial in F, $\partial[P(x)] = n$. The extension $E = F(\alpha)$, obtained by adjoining a root α of $P(x)$ to F, possesses the generators*

$$1, \alpha, \alpha^2, \ldots, \alpha^{n-1}.$$

Indeed, every element of $F(\alpha)$ can be written in the form

$$\theta = c_0 + c_1\alpha + \cdots + c_{n-1}\alpha^{n-1},$$

where $c_i \in F$. These generators are linearly independent, for otherwise α would be the root of a polynomial of degree lower than n—but this is impossible (Lemma 4.5, p. 40). It follows that they form a basis and therefore $(E/F) = n$.

THEOREM 5.10 *Let F be a ground field, E an extension of F, Ω an extension of E; $\Omega \supset E \supset F$. It follows that*

$$(\Omega/E)(E/F) = (\Omega/F).$$

PROOF: Consider first the cases where one of the factors is infinite:

(a) Assume (Ω/E) is infinite. In that case it is possible to find as many elements as we please, all linearly independent with respect to E. The same elements are manifestly independent with respect to F since $E \supset F$. It follows that (Ω/F) is infinite.

(b) Assume (E/F) is infinite. It is then possible to choose in E as many elements as we please, all linearly independent with respect to F. But these are also elements of Ω and consequently (Ω/F) is infinite.

In these cases there is no product $(\Omega/E)(E/F)$ in the proper sense, but we agree to define it so as to include (a) and (b) in the general statement of the theorem.

Assume now that $(E/F) = n$; $\omega_1, \omega_2, \ldots, \omega_n$ is a basis of E/F.[6] Similarly, let $(\Omega/E) = m$, and let $\Omega_1, \Omega_2, \ldots, \Omega_m$ be a basis of Ω/E. Every $\theta \in \Omega$ can be expressed as a linear combination

$$\theta = \alpha_1\Omega + \alpha_2\Omega + \cdots + \alpha_m\Omega_m,$$

[6]Read: "E over F" for "E/F."

where $\alpha_i \in E$. Each α_i is likewise expressible as a linear combination over F;

$$\alpha_i = a_{i1}\omega_1 + a_{i2}\omega_2 + \cdots + a_{in}\omega_n, \quad a_{ij} \in F.$$

Combining these results, we obtain

$$\theta = \sum_{i=1}^{m}\sum_{j=1}^{n} a_{ij}\omega_j\Omega_i;$$

i.e., the elements $\omega_j\Omega_i$ are nm generators of Ω with respect to F. We have only to prove that they constitute a basis of Ω/F. For the proof that the $\omega_i\Omega_j$ are linearly independent, set

$$a_{11}\omega_1\Omega_1 + a_{12}\omega_2\Omega_1 + \cdots + a_{m1}\omega_1\Omega_m + \cdots + a_{mn}\omega_n\Omega_m = 0.$$

Rewriting, we obtain

$$(a_{11}\omega_1 + \cdots + a_{1n}\omega_n)\Omega_1 + \cdots + (a_{m1}\omega_1 + \cdots + a_{mn}\omega_n)\Omega_m = 0$$

where the coefficients of the Ω_i are now certain elements of E. Since the Ω_i are linearly independent with respect to E, these coefficients must all be zero, i.e.,

$$a_{i1}\omega_1 + a_{i2}\omega_2 + \cdots + a_{in}\omega_n = 0, \quad i = 1, 2, \ldots, m.$$

But the ω_i are independent with respect to F; we conclude that each $a_{ij} = 0$ and the proof is complete. □

COROLLARY *If* $\Omega \supset E \supset F$ *and* $(E/F) = (\Omega/F)$, *then* $\Omega = E$.

This is a direct consequence of the following:

PROPOSITION

$$(E/F) = 1 \Rightarrow E = F.$$

PROOF: If the degree of E/F is one, then E is generated by any single element which is independent (i.e., nonzero). But 1 is independent; therefore every element of E is in F. Consequently $E = F$. □

This corollary can also be obtained as a special case of the corollary to Theorem 2.5 (p. 17).

5.6. Group Characters

Given a field E and a multiplicative group G, then a function $\lambda(x)$ which takes on values in E for arguments in G is called a *character* provided:

(a) $\lambda(x) \neq 0$ for some $x \in G$,
(b) $\lambda(xy) = \lambda(x)\lambda(y)$.

It is easy to establish that $\lambda(x) \neq 0$ for any element of G. From (a) there is a $c \in G$ such that $\lambda(c) \neq 0$. If for some $a \in G$, $\lambda(a) = 0$, then

$$\lambda(c) = \lambda(a)\lambda(a^{-1}c) = 0,$$

a contradiction.

LEMMA 5.11 *Let $\lambda_1(x), \lambda_2(x), \ldots, \lambda_n(x)$ be n distinct characters of G with values in E. Then if a linear combination*

$$c_1\lambda_1(x) + c_2\lambda_2(x) + \cdots + c_n\lambda_n(x) = 0$$

for all $x \in G$, it follows that $c_i = 0$, $i = 1, 2, \ldots, n$.

PROOF: Let us assume the contrary, that there are nontrivial linear relations among the λ_i. Select one of these for which the number of nonzero coefficients is least, say

$$c_1\lambda_1(x) + c_2\lambda_2(x) + \cdots + c_r\lambda_r(x) = 0 \qquad [1]$$

where $c_i \neq 0$, $i = 1, 2, \ldots, r$. Evidently $r \neq 1$, for $c_1\lambda_1(x) = 0$ implies that $c_1 = 0$ since $\lambda_1(x)$ is never zero. Therefore $r > 1$.

The relation [1] is assumed to be true for all $x \in G$. Consequently, it must be true if we substitute for x any argument in G. Replacing x by ax where $a \in G$, we find

$$c_1\lambda_1(a)\lambda_1(x) + c_2\lambda_2(a)\lambda_2(x) + \cdots + c_r\lambda_r(a)\lambda_r(x) = 0. \qquad [2]$$

Multiply by $\lambda_r(a)$ in [1] and subtract the result from [2]. This yields the relation

$$c_1[\lambda_1(a) - \lambda_r(a)]\lambda_1(x) + \cdots + c_{r-1}[\lambda_{r-1}(a) - \lambda_r(a)]\lambda_{r-1}(x) = 0 \qquad [3]$$

which is shorter than [1]. If it can be shown that not all these coefficients are zero, then this result contradicts the assumption that r is least and we are through.

We have assumed that $\lambda_1(x)$ and $\lambda_r(x)$ are distinct functions. Hence there is an $a \in G$ for which $\lambda_1(x) \neq \lambda_r(a)$. Let this be the a we have chosen above. In that case $c_1[\lambda_1(a) - \lambda_r(a)] \neq 0$ and [3] is a nontrivial relation which is shorter than [1]. □

REMARK. The symbol $a \to \bar{a}$ for a mapping is replaced by a functional notation; in particular, the image of a field element a through an automorphism σ is denoted by $\sigma(a)$. Clearly

$$\sigma(1) = 1 \neq 0$$

and $\sigma(xy) = \sigma(x)\sigma(y)$. Therefore σ is a character for the multiplicative group consisting of the nonzero elements of the field. Lemma 5.11 provides us with the important

THEOREM 5.12 *Let E be a field, and $\omega_1, \omega_2, \ldots, \omega_n$ distinct automorphisms of E. Then if*

$$c_1\omega_1(x) + c_2\omega_2(x) + \cdots + c_n\omega_n(x) = 0$$

for all $x \in E$, it follows that $c_i = 0$, $i = 1, 2, \ldots, n$.

LEMMA 5.13 *The set F of all elements of E which remain fixed under the automorphisms σ_i (i.e., the set consisting of all $a \in E$ such that $\sigma_i(a) = a$) is a subfield of E.*

PROOF: It is only necessary to show closure with respect to addition, subtraction, multiplication, and division. We have

$$\left.\begin{array}{r} a \in F \\ b \in F \end{array}\right\} \Rightarrow \left.\begin{array}{r} \sigma_i(a) = a \\ \sigma_i(b) = b \end{array}\right\} \Rightarrow \sigma_i(a \pm b) = a \pm b \Rightarrow a \pm b \in F.$$

Similarly, we perceive that $ab \in F$ and, if $b \neq 0$, $a/b \in F$. □

If F is the fixed field under a set of automorphisms of E, what is the degree of E/F? An exact answer cannot be given unless we assume something further about the automorphisms. However, we shall prove

THEOREM 5.14 *If F is the field consisting of the elements fixed under n distinct automorphisms $\sigma_1, \sigma_2, \ldots, \sigma_n$ of E, then $(E/F) \geq n$.*

REMARK. It may very well happen that the degree of E over the field F fixed under only one automorphism is already infinite. For example:

EXERCISE 4. Let $E = F(x)$, the field of rational functions over F.[7] (We shall hereafter denote the set of polynomials over F by $F[x]$.) Show that the mapping $f(x) \to f(x+1)$ is an automorphism of the field E and prove further that the fixed elements under this automorphism are the constants (i.e., the elements of F).

PROOF OF THEOREM 5.14: Assume $(E/F) = r < n$ and let $\omega_1, \omega_2, \ldots, \omega_r$ be a basis of E/F. Accordingly there are $c_i \in F$ for each $\theta \in E$ such that

[1] $$\theta = c_1\omega_1 + c_2\omega_2 + \cdots + c_r\omega_r.$$

The system

[2] $$\begin{cases} \xi_1\sigma_1(\omega_1) + \xi_2\sigma_2(\omega_1) + \cdots + \xi_n\sigma_n(\omega_1) = 0 \\ \vdots \\ \xi_1\sigma_1(\omega_r) + \xi_2\sigma_2(\omega_r) + \cdots + \xi_n\sigma_n(\omega_r) = 0 \end{cases}$$

of r linear homogeneous equations in n unknowns, $n > r$, has a nontrivial solution $\xi_1, \xi_2, \ldots, \xi_n$ in E. Multiply the ith equation in [2] by $c_i \in F$. Since c_i is in F, $\sigma_j(c_i) = c_i$, $j = 1, 2, \ldots, n$. Thus we obtain

[3] $$\begin{cases} \xi_1\sigma_1(c_1\omega_1) + \cdots + \xi_n\sigma_n(c_1\omega_1) = 0 \\ \vdots \\ \xi_1\sigma_1(c_r\omega_r) + \cdots + \xi_n\sigma_n(c_r\omega_r) = 0. \end{cases}$$

By adding the left sides in [3] we find

$$\xi_1\sigma_1(\theta) + \xi_2\sigma_2(\theta) + \cdots + \xi_n\sigma_n(\theta) = 0$$

where $\xi_1, \xi_2, \ldots, \xi_n$ are not all zero and θ, given by [1], may be any element of E. But this is contrary to the result of Theorem 5.12 and therefore the hypothesis $r < n$ is inadmissible. □

[7]The rational functions over F are the symbolic quotients of polynomials $\phi(x)/\psi(x)$, $\psi(x) \neq 0$, which are defined to equate, add, and multiply in the same manner as fractions.

EXAMPLE. Take $E = R(x)$, the field of rational functions over the field R of rational numbers. Consider the result of

EXERCISE 5. Show that all six mappings

$$f(x) \to f(x),\ f\left(\frac{1}{x}\right),\ f(1-x),\ f\left(\frac{1}{1-x}\right),\ f\left(\frac{x}{x-1}\right),\ f\left(\frac{x-1}{x}\right)$$

are automorphisms of $R(x)$. (Note that these automorphisms form a group.)[8]

If F is the field consisting of the elements which remain fixed under these automorphisms, then $(E/F) \geq 6$. What are the elements of F? It is easy to verify, in particular, that

$$J(x) = \frac{(x^2 - x + 1)^3}{x^2(x-1)^2}$$

is in F. Clearly all rational functions of $J(x)$ are in F. If we denote the field consisting of the rational functions of $J(x)$ by F_0, we have $E \supset F \supset F_0$ and therefore $(E/F_0) \geq 6$.

Now, clearly, $E = F_0(x)$; i.e., E can be obtained from F_0 by adjoining x. Manifestly, x satisfies a sixth-degree equation over F_0, namely $x^2(x-1)^2 J = (x^2 - x + 1)^3$. Consequently, $E/F_0 \geq 6$ (Lemma 5.9). It follows that $(E/F_0) = 6$. Furthermore, the equation for x is irreducible since by Lemma 5.9 it cannot satisfy an irreducible equation of lower degree. Since $F_0 = F$ (cf. the proposition on p. 60) we have found all rational functions which remain fixed under these automorphisms.

Similarly, for the fixed field F of the subgroup of automorphisms

(a) $f(x) \to f(x), f(1/x), (E/F) = 2$, and F consists of the rational functions of $J = x + 1/x$.

(b) $f(x) \to f(x), f(1-x), (E/F) = 2$, and F consists of the rational functions of $J = x(1-x)$.

EXERCISE 6. Determine the nature of the elements fixed under the subgroup of automorphisms of order 3: $f(x) \to f(x), f(\frac{1}{1-x}), f(\frac{x-1}{x})$.

5.7. Automorphic Groups of a Field

PROPOSITION *The set of all automorphisms of a field or, for that matter,* any *mathematical system S is a group.*

REMARK. The *product* of two automorphisms $\sigma(x)$ and $\tau(x)$ is defined to be the mapping $\sigma(\tau(x))$ of S into itself. We denote this product briefly by $\sigma\tau$.

PROOF:

(a) Closure. Since the truth value of a relation among the elements of S is preserved by automorphisms, it is preserved by their products. Any product $\sigma\tau$ is a 1-1 mapping; for since the argument $\tau(x)$ runs through all the elements of S, so must $\sigma(\tau(x))$ and, also, since the images of distinct elements of S by τ and σ

[8]Cf. Exercise 1, p. 1.

are distinct, the images by $\sigma\tau$ are distinct. We conclude that the product of two automorphisms is again an automorphism.

(b) The associative law has already been proved for functions in general.[9]

(c) There is an identity I such that $I\sigma = \sigma$. This is the automorphism $I(x) = x$ which maps each element of S onto itself.

(d) To each automorphism σ of S there is an inverse σ^{-1} which associates to each element an image by the inverse mapping; i.e., if $\sigma(x) = y$ we define σ^{-1} by $\sigma^{-1}(y) = x$. Thus

$$\sigma^{-1}(\sigma(x)) = x = I(x).$$

The inverse mapping is clearly an automorphism since a relation is true or false for images through σ according to the validity of the same relation for their antecedents. It follows that the inverse mapping preserves the truth value of relations in S.

□

Consider a field E together with a finite number of its automorphisms $\sigma_1, \sigma_2, \ldots, \sigma_n$, and let F be the subfield consisting of the fixed elements under the σ_i. We have proved $(E/F) \geq n$. If it is possible to find more automorphisms of E for which F remains fixed, then this result may be improved. We may immediately improve the result by including all possible products of the σ_i; for, if $a \in F$ is fixed under two automorphisms σ and τ, it is fixed under the product $\sigma\tau$, that is,

$$\left.\begin{array}{l}\sigma(a) = a\\ \tau(a) = a\end{array}\right\} \Rightarrow \sigma(\tau(a)) = a.$$

There are two possibilities:

(1) We may in this manner be able to obtain any number of automorphisms we please. In that case $(E/R) = \infty$.

EXAMPLE. Let $R(x)$ be the field of rational functions over the field R of rational numbers. From the automorphism $\sigma\tau$, $f(x) \to f(x+1)$, we obtain the automorphisms

$$\begin{aligned}\sigma[f(x)] &= f(x+1),\\ \sigma^2[f(x)] &= f(x+2),\\ &\vdots\\ \sigma^n[f(x)] &= f(x+n),\\ &\vdots\end{aligned}$$

which are all distinct. The degree of $R(x)$ over R is therefore infinite.[10]

(2) On the other hand, the set of all possible products of the σ_i may be finite. In that event they form a group from the result of

---

[9]See p. 2.

[10]Cf. Exercise 4, p. 62.

EXERCISE 7. Prove that a finite subset of a group is a subgroup provided only that it be closed with respect to multiplication.

In any case we may assume that the set of automorphisms σ_i is a group G. For if not, we may append all possible products of the σ_i; if the identity is not among these it also is added; furthermore, the inverses may be adjoined since they too leave F fixed. Once we arrive at a group it is impossible by this method to improve our information about the degree of E/F any further. In fact, we have obtained a complete result:

THEOREM 5.15 *Let E be a field, and $\sigma_1, \sigma_2, \dots, \sigma_n$ a group G of automorphisms of E. If F is the fixed field of this group, then $(E/F) = n$ precisely.*

REMARK. Consider the set consisting of the automorphisms

$$\sigma_i\sigma_1, \sigma_i\sigma_2, \dots, \sigma_i\sigma_n.$$

Since these are n distinct elements of G, it follows that they are merely the σ's in another arrangement. Consequently, if

$$a = \sigma_1(\theta) + \sigma_2(\theta) + \cdots + \sigma_n(\theta) \tag{1}$$

where $\theta \in E$, we conclude that $a \in F$; i.e., a is a fixed point. For we may write

$$\sigma_i(a) = \sigma_i \sum_{k=1}^{n} \sigma_k(\theta) = \sum_{k=1}^{n} \sigma_i\sigma_k(\theta) = \sum_{k=1}^{n} \sigma_k(\theta) = a.$$

Furthermore, all elements of F may be described in the form [1]. For by Theorem 5.12 there is a $\theta \in E$ for which $a \neq 0$. Therefore in order to express any $b \in F$ in the form [1], we have only to multiply θ by b/a.

PROOF: Let $\alpha_1, \alpha_2, \dots, \alpha_m$ be any m elements of E. The theorem is proved by showing that if $m > n$ the α's are linearly dependent and hence $(E/F) \leq n$. From Theorem 5.14 it then follows that $(E/F) = n$ and consequently that the group G contains *all* automorphisms which leave F fixed.

Consider the system

$$\begin{cases} x_1\sigma_1(\alpha_1) + x_2\sigma_1(\alpha_2) + \cdots + x_m\sigma_1(\alpha_m) = 0 \\ \vdots \\ x_1\sigma_n(\alpha_1) + x_2\sigma_n(\alpha_2) + \cdots + x_m\sigma_n(\alpha_m) = 0 \end{cases} \tag{2}$$

of n linear equations in m unknowns, $m > n$. This system has a nontrivial solution $x_1, x_2, \dots, x_m$ (Theorem 2.1, p. 13) with say $x_1 \neq 0$. Clearly $\lambda x_1, \lambda x_2, \dots, \lambda x_m$ is also a solution for any $\lambda \in E$. Select λ so that $\lambda x_1 = \theta$ where θ gives a nonzero $a \in F$ by [1]. We may assume that $x_1 = \theta$ in our solution.

Applying σ_i to the system [2] we obtain

$$\sigma_i(x_1)\sigma_i\sigma_k(\alpha_1) + \sigma_i(x_2)\sigma_i\sigma_k(\alpha_2) + \cdots + \sigma_i(x_m)\sigma_i\sigma_k(\alpha_m) = 0,$$
$$k = 1, 2, \dots, n.$$

Since $\sigma_i\sigma_k$ $(k = 1, 2, \dots, n)$ are all n automorphisms, this is the same system as [2] where $x_1, x_2, \dots, x_m$ are replaced by $\sigma_i(x_1), \sigma_i(x_2), \dots, \sigma_i(x_m)$, respectively.

Consequently, $\sigma_i(x_1), \ldots, \sigma_i(x_m)$ is also a solution of [2]. Furthermore, for any system of homogeneous linear equations the sum of two solutions is again a solution. It follows that $x_1', x_2', \ldots, x_m'$ where

$$x_j' = \sum_{i=1}^{n} \sigma_i(x_j), \quad j = 1, 2, \ldots, m,$$

is also a solution. The solution is nontrivial, for putting $x_1 = \theta$ we obtain $x_1' = a \neq 0$. Now, identity appears among the σ_i since they form a group. Therefore one of the equations in [2] takes the form

$$x_1'\alpha_1' + x_2'\alpha_2 + \cdots + x_m'\alpha_m = 0$$

where $x_i' \in F$, $i = 1, 2, \ldots, m$. The α's are linearly dependent. □

THEOREM 5.16 *Let G be a finite group of automorphisms $\sigma_1, \sigma_2, \ldots, \sigma_n$ of the field E, and denote by F the fixed field of the σ_i. Then any element α of E is algebraic over F; i.e., α is the solution of a polynomial equation over F.*

PROOF: Consider $\sigma_1(\alpha), \sigma_2(\alpha), \ldots, \sigma_n(\alpha)$, the images of α through the elements of G. Pick out the $\alpha_i = \sigma_i(\alpha)$ which are distinct, say $\alpha_1, \alpha_2, \ldots, \alpha_r$, $r < n$. Manifestly α itself is one of the α_i since I is one of the σ_i. Now $\sigma_i\sigma_1(\alpha), \sigma_i\sigma_2(\alpha), \ldots, \sigma_i\sigma_r(\alpha)$ are all different since the images of different elements by the same automorphism are distinct. But these are part of the set consisting of $\sigma_i\sigma_1(\alpha), \sigma_i\sigma_2(\alpha), \ldots, \sigma_i\sigma_n(\alpha)$ and therefore are merely the distinct elements $\alpha_1, \alpha_2, \ldots, \alpha_r$ in another arrangement. Set

$$\phi(x) = \prod_{k=1}^{n}(x - \alpha_k).$$

It follows that

$$\sigma_i(\phi(x)) \prod_{k=1}^{r} \sigma_i(x - \alpha_k) = \prod_{k=1}^{r}(x - \sigma_i'(\alpha_k)) = \phi(x).$$

Since the coefficients of $\phi(x)$ are unchanged by the automorphisms in G, we conclude that they are elements of F. But $\phi(x)$ has the roots $\alpha_1, \alpha_2, \ldots, \alpha_r$ of which one is α. This is the desired conclusion. □

COROLLARY *The polynomial $\phi(x)$ over F is even irreducible.*

Let $f(x)$ be any polynomial over F with the root α, $f(\alpha) = 0$. The σ_i do not change the coefficients of $f(x)$. Therefore

$$\sigma_i(f(\alpha)) = f(\sigma_i(\alpha)) = f(\alpha_i) = 0$$

whence $f(x)$ has at least the roots $\alpha_1, \alpha_2, \ldots, \alpha_r$ and $\partial[f(x)] \geq r$. $\phi(x)$ is then the polynomial of least degree for α and is therefore irreducible (Lemma 4.4, p. 40). Thus for any $a \in E$ the method of this theorem provides an irreducible polynomial $\phi(x)$ over F which has α as a root.

We note that the polynomial for α does not have a multiple root and hence the case we have discussed (p. 55 ff.) cannot occur under the conditions for Theorem

5.16. If the roots of an irreducible polynomial are all simple, the polynomial is said to be *separable*. In general, any polynomial will be called *separable* if each of its irreducible factors is separable. When the roots of a polynomial are simple, it is certainly separable. However, this is not a necessary condition since, e.g., $[\phi(x)]^2$ is separable. It will be recalled that Theorem 5.5 on fixed fields requires the use of separable polynomials.

EXAMPLE. Consider the polynomial $f(x) = x^4 - 2$ over the rational field R. Let us construct the splitting field E of $f(x)$. In the field of complex numbers $x^4 - 2$ splits into the factors

$$(x - \sqrt[4]{2})(x + \sqrt[4]{2})(x - i\sqrt[4]{2})(x + i\sqrt[4]{2}).$$

We conclude that $E = R(\sqrt[4]{2}, i\sqrt[4]{2}) = R(\sqrt[4]{2}), i)$. What degree is the splitting field? We have

$$(E/R) = (E/R(\sqrt[4]{2}))(R(\sqrt[4]{2})/R).$$

$(E/R(\sqrt[4]{2})) = 2$ since i satisfies the irreducible equation $x^2+1=0$ of second degree over the field $R(\sqrt[4]{2})$. Now $\sqrt{2}$ is irrational; for suppose there were integers m, n such that $\sqrt{2} = m/n$ where m and n are relatively prime. We may then write $2 = m^2/n^2$. Since the quotient is an integer, we must have $n^2 = 1$ whence $m^2 = 2$. But this implies that $\sqrt{2}$ is an integer, which is clearly not true. Thus $\sqrt{2}$ cannot be rational. It follows that $x^4 - 2$ cannot be the product of two quadratic factors over R. Consequently,

$$(R(\sqrt[4]{2})/R) = 4 \quad \text{and} \quad (E/R) = 8.$$

What are all automorphisms of E which leave R fixed? We have shown (p. 44) that the rational numbers always remain fixed. Hence we have only to find all automorphisms of E. Since $f(x)$ is separable, Theorem 5.5 tells us that no elements but those of R remain fixed. But the set of all automorphisms is a group and R is the fixed field of this group. We conclude that there are exactly eight automorphisms. It is not difficult to write these down:

	$\sqrt[4]{2}$	i
I	$\sqrt[4]{2}$	i
σ	$i\sqrt[4]{2}$	i
σ^2	$-\sqrt[4]{2}$	i
σ^3	$-i\sqrt[4]{2}$	i
τ	$\sqrt[4]{2}$	$-i$
$\sigma\tau$	$i\sqrt[4]{2}$	$-i$
$\sigma^2\tau$	$-\sqrt[4]{2}$	$-i$
$\sigma^3\tau$	$-i\sqrt[4]{2}$	$-i$

EXERCISE 8. Demonstrate that this group of eight automorphisms is isomorphic to the group of symmetries of a square. Denote by σ the automorphism $(\sqrt[4]{2}, i) \to (i\sqrt[4]{2}, i)$. The powers of σ form a cyclic subgroup of order 4. If we denote by τ the automorphism $(\sqrt[4]{2}, i) \to (\sqrt[4]{2}, -i)$, then we can describe all products in terms of σ and τ by means of the rules $\sigma^4 = I$, $\tau^2 = I$, $\tau\sigma = \sigma^3\tau$, $\tau\sigma^2 = \sigma^2\tau$, $\tau\sigma^3 = \sigma\tau$.

Let us determine all subgroups. These are classified as follows:

(1) Order 8 G_8: the entire group

(2) Order 4 (a) the cyclic group

$$C_4:\ I, \sigma, \sigma^2, \sigma^3$$

(b) the four groups (all elements have period 2)

$$G_{41}: I, \sigma^2, \tau, \sigma^2\tau$$
$$G_{42}: I, \sigma^2, \sigma\tau, \sigma^3\tau$$

(3) Order 2 $G_{21}:\ I, \sigma^2$
$G_{22}:\ I, \tau$
$G_{23}:\ I, \sigma\tau$
$G_{24}:\ I, \sigma^2\tau$
$G_{25}:\ I, \sigma^3\tau$

(4) Order 1 Consists of the element I alone.

The relations among the subgroups are indicated by the scheme

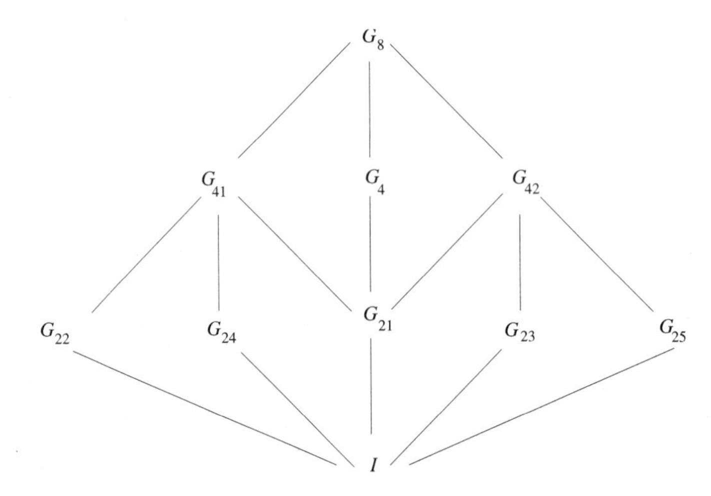

The subfields of E which correspond to these subgroups are interrelated in the same manner except, however, that the larger field corresponds to the smaller group. Thus if N is the order of a subgroup and F is the corresponding field, then $(E/F) = N$ and hence $(F/R) = 8/N$.

A group of automorphisms of a field determines a subfield consisting of the elements left fixed. The converse is not true; not every subfield can be described as the fixed field of a group of automorphisms. For example, the field $R(\sqrt[3]{2})$ has only the identity automorphism. Consequently, R cannot be described as the fixed field of some group of automorphisms of $R(\sqrt[3]{2})$. What is the distinguishing characteristic of the fixed field of an automorphic group?

5.8. Fundamental Theorem of Galois Theory

If F is the fixed field of a finite group of automorphisms of the field E, we say E is *normal* over F or F is normal under E and write E/F is normal.

THEOREM 5.17 *E/F is normal if and only if E is the splitting field of a separable polynomial over F.*

PROOF: *Sufficiency.* Assume E is the splitting field of a separable polynomial $f(x)$ over F. Then by Theorem 5.5, F is the fixed field under the group of all automorphisms which leave at least every element of F fixed. Therefore E/F is normal.

Necessity. Assume E/F is normal, $(E/F) = n$. Then there is a basis $\omega_1, \omega_2, \ldots, \omega_n$ of E/F and E is obtained from F by adjoining the ω_i: $E = F(\omega_1, \omega_2, \ldots, \omega_n)$. Since (E/F) is finite, each ω_i is a root of an irreducible separable polynomial $p_i(x)$ over F (Theorem 5.16, p. 66). The polynomial

$$f(x) = p_1(x)p_2(x)\cdots p_n(x)$$

splits in E since each factor splits in E. Moreover, among the roots are $\omega_1, \omega_2, \ldots, \omega_n$. Hence no smaller field than E can possibly be the splitting field of $f(x)$. The proof is complete. □

COROLLARY 5.18 *If E/F is normal and if Ω is any field intermediate between E and F, $F \subset \Omega \subset E$, then E/Ω is normal.*

PROOF: E is the splitting field of a polynomial $f(x)$ over F and consequently is the splitting field of the same polynomial over Ω. □

COROLLARY 5.19 *If G is the group of E/F (i.e., F is the fixed field under the group G of automorphisms of E), then there is a 1-1 correspondence between the subgroups of G and the subfields of E which contain F:*
$S \subset G \Leftrightarrow \exists\Omega$ such that $F \subset \Omega \subset E$ where S is the group of E/Ω.

The proof is obvious.

EXAMPLE. (Cf. with example on p. 67.) Consider the field $E = R(\sqrt[4]{2}, i)$ over the field of rational numbers. Each intermediate field between E and R corresponds to a subgroup of G. Thus there are three subfields of degree 2 corresponding to the subgroups of order 4 and five subfields of degree 4 corresponding to the subgroups of order 2. It is easy to find the fields of degree 2: $R(i) \leftrightarrow G_4$, $R(\sqrt{2}) \leftrightarrow G_{41}$, $R(i\sqrt{2}) \leftrightarrow G_{42}$. Of degree 4 we have the fields $R(i, \sqrt{2}) \leftrightarrow G_{21}$, $R(\sqrt[4]{2}) \leftrightarrow G_{22}$, $R(i\sqrt[4]{2}) \leftrightarrow G_{24}$. However, it is not always easy to tell on sight which field corresponds to a given group. What are the fixed fields of G_{23} and G_{25}?

The fixed field of G_{23} consists of elements which are not changed by $\sigma\tau$, $(\sqrt[4]{2}, i) \to (i\sqrt[4]{2}, -i)$. The general element of E may be put in the form

$$\begin{aligned}\theta = c_0 + c_1\sqrt[4]{2} + c_2\sqrt{2} + c_3(\sqrt[4]{2})^3 + c_4 i\\ + c_5 i\sqrt[4]{2} + c_6 i\sqrt{2} + c_7 i(\sqrt[4]{2})^3,\end{aligned}$$

whence

$$(\theta) = c_0 + c_1 i\sqrt[4]{2} + c_2(-\sqrt{2}) + c_3(-i\sqrt[4]{2^3}) + c_4(-i) + c_5\sqrt[4]{2} + c_6 i\sqrt{2} + c_7(-\sqrt[4]{2^3}).$$

If θ remains unchanged

$$c_0 \text{ arbitrary}, \quad c_2 = 0, \quad c_3 = -c_7, \quad c_4 = 0, \quad c_6 \text{ arbitrary}.$$

Namely,

$$\begin{aligned}\theta &= c_0(1+i)\sqrt[4]{2} + c_6 i\sqrt{2} + c_3(1-i)\sqrt[4]{2^3} \\ &= c_0(1+i)\sqrt[4]{2} + \frac{c_6}{2}\left[(1+i)\sqrt[4]{2}\right]^2 - \frac{c_3}{2}\left[(1+i)\sqrt[4]{2}\right]^3\end{aligned}$$

whence $(1+i)\sqrt[4]{2} = \sqrt[4]{-8}$ generates the field.

The interrelations of the subfields and their groups is given by the following scheme:

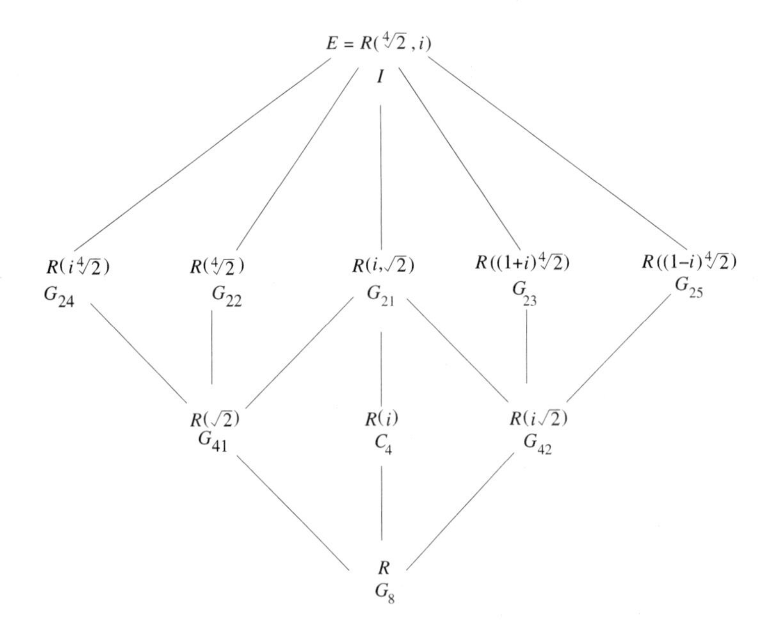

EXAMPLE. Consider the field $R(\sqrt[3]{2})$ which is of degree 3 over R and yet possesses no other automorphism than the identity. The number $\sqrt[3]{2}$ satisfies the irreducible equation $x^3 - 2 = 0$ over R. From Theorem 5.17 it is clear that we must go to the splitting field of this equation in order to obtain a better understanding. The other solutions of the equation are $\omega\sqrt[3]{2}$ and $\omega^2\sqrt[3]{2}$ where ω is a root of the irreducible polynomial $x^2 + x + 1 = 0$ over $R(\sqrt[3]{2})$. Thus we write

$$x^3 - 2 = (x - \sqrt[3]{2})(x - \omega\sqrt[3]{2})(x - \omega^2\sqrt[3]{2})$$

whence $E = R(\sqrt[3]{2}, \omega)$. It follows that $(E/R) = 6$. There are then exactly six automorphisms of E, for E is a splitting field and therefore normal. These automorphisms are determined by the manner in which they transform the roots of the above equations. The root $\sqrt[3]{2}$ can have only three possible images, ω only two. There are six possible combinations and since there are six automorphisms all combinations occur. The automorphic group is given by the table:

	I	σ	σ^2	τ	$\sigma\tau$	$\sigma^2\tau$
$\sqrt[3]{2}$	$\sqrt[3]{2}$	$\omega\sqrt[3]{2}$	$\omega^2\sqrt[3]{2}$	$\sqrt[3]{2}$	$\omega\sqrt[3]{2}$	$\omega^2\sqrt[3]{2}$
ω	ω	ω	ω	ω^2	ω^2	ω^2

The group multiplication follows the rules

$$\sigma^3 = I,\ \tau^2 = I;\ \ \sigma\tau = \sigma^2\tau,\ \tau\sigma^2 = \sigma\tau.$$

It is easy to verify that this is the group of symmetries of the triangle (cf. Exercise 1, p. 1).

The group of E/R possesses one subgroup of order 3, G_3: I, σ, σ^2; and three subgroups of order 2,

$$G_{21}\colon I, \tau; \quad \sigma_{22}\colon I, \sigma\tau; \quad G_{23}\colon I, \sigma^2\tau.$$

The subgroups correspond to the one quadratic field $R(\omega)$ and three cubic fields, $R(\sqrt[3]{2})$, $R(\omega\sqrt[3]{2})$, and $R(\omega^2\sqrt[3]{2})$, respectively.

EXERCISE 9. Determine the automorphisms of the polynomial for which $\sqrt[5]{2}$ is a root.

Consider the general problem of an arbitrary irreducible cubic $P(x)$ over a field F with distinct roots $\alpha_1, \alpha_2, \alpha_3$. The splitting field

$$E = F(\alpha_1, \alpha_2, \alpha_3)$$

is in general of degree six. If, however, the equation already splits in $F(\alpha_1)$, then $(E/F) = 3$ (Lemma 5.9). In this special case the group of E/F being of prime order is therefore cyclic and consists of the powers of one element I, σ, σ^2. We conclude that σ must permute the roots cyclically; for all other permutations would leave one root fixed, and hence would consist of at most the transposition of two roots and be therefore of period 2. Thus σ is represented either by the permutation $\binom{123}{231}$ or $\binom{123}{312}$ and there are no other possibilities.

EXERCISE 10. Find the irreducible polynomial over R with root $2\cos\frac{2\pi}{7}$. Show that this is a cubic of the above special type. (*Hint*: Use seventh roots of unity.)

On the other hand, suppose $F(\alpha_1)$ is not the splitting field. Since $x-\alpha_1$ may be factored out in $F(\alpha_1)$, we must have $(E/F(\alpha_1)) = 2$ and consequently $(E/F) = 6$. There must then be six automorphisms of E which leave F fixed. These are determined by the way they permute the roots. Since there are six permutations of three elements, all permutations are possible. Thus we have shown that the group of automorphisms of the splitting field of an irreducible cubic is either the triangle group or the cyclic group of order 3. No other cases occur. We shall see that this result implies the solvability of the general cubic by radicals.

For the irreducible equation of fourth degree there are 24 possible permutations of the roots. In most cases the splitting field will actually be of degree 24. The special cases correspond to subgroups of the permutation group of four objects. In order to know all possibilities, take the group of 24 elements and find all possible subgroups corresponding to the irreducible case. In general, the analysis of the general equation of nth degree involves the group of permutations of n objects. The splitting field is most often of the highest possible degree—$n!$. It will appear later that this method of treating the solution enables us to tell whether or not any given equation is solvable in terms of radicals.

THEOREM 5.20 *Let U be a field containing*

(1) *F, the ground field.*
(2) *E, the splitting field of any polynomial $f(x)$ (not necessarily separable) over F.*
(3) *Ω, a field intermediate between E and F, $E \supset \Omega \supset F$.*
(4) *Ω', an extension field of F which is isomorphic to Ω in a mapping which leaves the elements of F fixed.*

It follows that $\Omega' \subset E$ and that the isomorphism between Ω and Ω' is contained in some automorphism of E.

Thus we see that it is possible to generalize our arguments to all polynomials.

PROOF: Denote the roots of $f(x)$ by $\alpha_1, \alpha_2, \ldots, \alpha_n$. The splitting field E of $f(x)$ is obtained by adjoining the roots:

$$E = F(\alpha_1, \alpha_2, \ldots, \alpha_n).$$

Since $E \supset \Omega \supset F$, E is also the splitting field of $f(x)$ over Ω. Furthermore, $f(x)$ is a polynomial in Ω' and the splitting field of $f(x)$ over Ω' is some field in U

$$E' = \Omega'(\alpha_1, \alpha_2, \ldots, \alpha_n).$$

By Theorem 5.3 the isomorphism between Ω and Ω' can be extended to E and E'. Now let ω be any element of Ω, ω' its image in Ω'. Since $\omega \in E$ it can be written in the form

$$\omega = \phi(\alpha_1, \alpha_2, \ldots, \alpha_n)$$

where ϕ is a polynomial with coefficients in F. In the isomorphism between E and E', ω can only be mapped on

$$\omega' = \phi(\alpha_1', \alpha_2', \ldots, \alpha_n')$$

where the α_i' are the images of the α_i. Since the coefficients of $f(x)$ lie in the fixed field F, the α_i' are again roots of $f(x)$ and are therefore just the α_i in another permutation. It follows that $\omega' \in E$. We conclude that $\Omega' \subset E$ and that $E' = E$ since both are splitting fields of $f(x)$ over Ω'. Thus the mapping of Ω onto Ω' is contained in an automorphism of E. The theorem is proved. □

Assume E/F normal and let Ω and Ω' be two fields intermediate between E and F which are isomorphic under some mapping which leaves F fixed. We know that every such isomorphism of Ω and Ω' is contained in an automorphism of E.

Conversely, any automorphism of E which leaves F fixed clearly maps Ω upon some intermediate field Ω'. Let us denote the group of E/F by G, that of E/Ω by S. It is natural to ask how many distinct mappings of Ω onto isomorphic fields are obtained through elements of G.

Choose $\sigma, \tau \in G$. We seek conditions that they both produce the same mapping of Ω. Now

$$\sigma(x) = \tau(x) \Leftrightarrow \tau^{-1}\sigma(x) = x.$$

Thus, if σ and τ map Ω in the same way, then Ω remains fixed in the mapping $\tau^{-1}\sigma$. It follows that $\tau^{-1}\sigma \in S$, the group of E/Ω, or $\sigma \in \tau S$. We have proved if $\sigma(x) = \tau(x)$ for all $x \in \Omega$ that σ and τ belong to the same left coset σS in G where S is the group of E/Ω.

The number of *different* mappings of Ω through elements of G is simply the number of left cosets of S, the so-called *index* of the subgroup S. If j is the index of S we have (Theorem 1.4, p. 5)

$$N = jn$$

where N is the order of G, and n the order of S. From $(E/F) = N$, $(E/\Omega) = n$, we see that $(\Omega/F) = j$.

Let Ω' be an image of Ω through an element of G: $\sigma(\Omega) = \Omega'$. What is the subgroup of E/Ω'? It contains all τ for which $\tau(x) = x$ whenever $x \in \Omega'$, i.e., all τ for which $\tau(\sigma(y)) = \sigma(y)$ when $y \in \Omega$. But

$$\tau(\sigma(y)) = \sigma(y) \Leftrightarrow \sigma^{-1}\tau\sigma(y) = y \Leftrightarrow \sigma^{-1}\tau\sigma \in S \Leftrightarrow \tau \in \sigma S\sigma^{-1}.$$

Thus the group of E/Ω' is $\sigma S\sigma^{-1}$, the so-called *conjugate* of S with respect to σ.

If for all $\sigma \in G$ we have

$$\sigma S\sigma^{-1} = S,$$

we then say S is an invariant subgroup of G. We now prove

LEMMA 5.21 *If E/F is normal and has the group G, a necessary and sufficient condition that an intermediate field Ω, $E \supset \Omega \supset F$, be normal with respect to F is that the group S of E/Ω be an invariant subgroup of G.*

PROOF: If Ω/F is normal, then from $(\Omega/F) = j$ we see that all j isomorphisms of Ω are automorphisms. Hence for all $\sigma \in G$ we have $\sigma(\Omega) = \Omega$ or $\sigma S\sigma^{-1} = S$; i.e., S is an invariant subgroup.

Conversely, if $\sigma S\sigma^{-1} = S$ for all σ, then $\sigma(\Omega) = \Omega$. The j isomorphisms of Ω are actually mappings of Ω onto itself. The fixed elements under the j automorphisms of Ω satisfy $\sigma(x) = x$ for all $\sigma \in G$ and consequently are contained in F. Since F is the fixed field of the group of automorphisms of Ω, it follows that Ω/F is normal. □

If Ω/F is normal, then every element of G provides suitable automorphisms of Ω. However, we have already shown that there are only j distinct mappings of Ω through elements of G. Thus the automorphisms of Ω/F are the group G with a new equivalence relation:

Two elements are equivalent if and only if they lie in the same coset

$$\sigma \equiv \tau \Leftrightarrow \sigma(x) = \tau(x) \quad \text{for all} \quad x \in \Omega \Leftrightarrow \sigma \in \tau S.$$

In this manner we define a new group G/S called the *factor group* of G with respect to S.

EXERCISE 11. Show that an equivalence relation which preserves multiplication in a group G,

$$a \equiv b, \quad c \equiv d \Rightarrow a \cdot c \equiv b \cdot d,$$

may be defined in terms of a unique invariant subgroup S so that two elements are equivalent if and only if they are contained in the same coset of S.

REMARK. The right and left cosets of an invariant subgroup with respect to a given element are the same and, conversely, if the right and left cosets of a subgroup with respect to any element are the same, then the subgroup is invariant, for we have

$$\sigma S \sigma^{-1} = S \Leftrightarrow \sigma S = S\sigma.$$

Let us determine all the subgroups and factor groups of any cyclic group G. Since the group is commutative, it follows by the remark above that every subgroup is invariant. If G is of order N, we may write its elements as follows:

$$\sigma, \sigma^2, \sigma^3, \ldots, \sigma^{N-1}, \sigma^N = 1.$$

Let S be any subgroup and let $r > 0$ be the smallest positive power such that $\sigma^r \in S$. For any $\sigma^s \in S$ we have $r|s$. First we may write $s = qr + p$ where $0 \leq p < r$. It follows that $\sigma^p = \sigma^s \sigma^{-qr} \in S$ and hence that $p = 0$. Thus $r|s$. Now $\sigma^N = 1 \in S$ and therefore $n = N/r$ is an integer. Clearly, S consists of the elements

$$1, \sigma^r, \sigma^{2r}, \ldots, \sigma^{(n-1)r}.$$

S is a cyclic group with generator σ^r and the order of S is n.

Conversely, for any divisor r of N, $N = rn$, there is a subgroup of order n generated by the element σ^r. We have shown that there are precisely as many subgroups as there are divisors of N.

What are the factor groups? We divide G into cosets of S

$$S = \{1, \sigma^r, \sigma^{2r}, \ldots, \sigma^{(n-1)r}\}$$

$$\sigma S = \{\sigma, \sigma^{r+1}, \sigma^{2r+1}, \ldots, \sigma^{(n-1)r+1}\}$$

$$\vdots$$

$$\sigma^{r-1} S = \{\sigma^{r-1}, \sigma^{2r-1}, \ldots, \sigma^{nr-1}\} = \sigma^{-1} S.$$

The cosets are better written

$$\sigma S, (\sigma S)^2, \ldots, (\sigma S)^{r-1}, (\sigma S)^r = S.$$

Hence the factor group G/S is a cyclic group of order r. In brief, for any cyclic group, all subgroups and all factor groups are cyclic.

As an example, suppose the group of E/F is cyclic of order 12. The hierarchy of fields normal with respect to F is best described by the diagram in Figure 5.1.

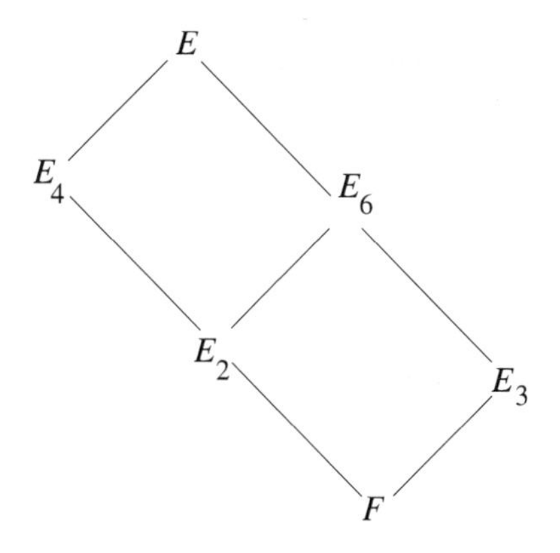

FIGURE 5.1

Thus to reach a cyclic extension of degree 12 from F we must make two quadratic extensions and a cubic extension.

5.9. Finite Fields

A finite field F cannot have characteristic zero for otherwise it would contain the infinite set of elements $1, 2, 3, \ldots$. By Lemma 5.6 (p. 54) the characteristic of F must then be a prime p. It does not hold, on the other hand, that every field of characteristic p is finite. Consider, for example, the field of rational functions $F(x)$ derived from the ring $F[x]$ of polynomials over F by forming all symbolic quotients of polynomials. More generally, we have

LEMMA 5.22 *If R is a commutative ring and contains no divisors of zero,*[11] *then it can be imbedded in a "quotient" field F consisting of the formal quotients a/b, $b \neq 0$.*

PROOF: We use the usual ordered pair definition as for rational numbers.

Equality: $a/b = c/d \Leftrightarrow ad = bc$

Addition: $a/b \pm c/d = (ad \pm bc)/bd$

Multiplication: $a/b \cdot c/d = a \cdot c/b \cdot d$.

The proof that equality preserves addition and multiplication is left to the reader. □

REMARK. F contains a ring isomorphic to R. If $a \in R$, then clearly $a \leftrightarrow a \cdot b/b, b \neq 0$.

EXERCISE 12. Let F be a field of characteristic 2. Consider the field $F(x^2)$ that consists of the rational functions in x^2 over F. $F(x^2) \subset F(x)$. Prove that $(F(x)/F(x^2)) = 2$. This is an example of a nonnormal extension of degree 2. If the characteristic of F were not 2, this quadratic extension would always be normal.

[11] Namely, $ab = 0 \Rightarrow a = 0$ or $b = 0$.

Let F be a finite field of characteristic p. F contains the field R_p of residue classes of the integers mod p

$$R_p = \{1, 2, 3, \ldots, p-1, p = 0\}.$$

Since the number of elements in F is finite, the degree $n = (F/R_p)$ is finite. Let $\omega_1, \omega_2, \ldots, \omega_n$ be a basis of F/R_p. Then every $\theta \in F$ may be uniquely described in the form

$$\theta = c_1\omega_1 + c_2\omega_2 + \cdots + c_n\omega_n$$

where the c_i are elements of R_p. The number of elements in F is therefore

$$q = p^n.$$

We have proved

LEMMA 5.23 *The number q of elements in a finite field F is the nth power of the characteristic where $n = (F/R_p)$.*

The $q-1$ nonzero elements of F form a multiplicative group of order $q-1$. Hence for all nonzero $\alpha \in F$

$$\alpha^{q-1} = 1. \tag{[1]}$$

Therefore, for all α in F we have

$$\alpha^q = \alpha, \tag{[2]}$$

a generalization of the Fermat theorem of arithmetic. It follows that the polynomial $x^q - x$ has q roots—the totality of elements of F. Since the degree of the polynomial is q, it can have no other roots. Hence

$$x^q - x = \prod_{\alpha \in F}(x - \alpha). \tag{[3]}$$

REMARK. This is equivalent to

$$x^q - x = x\prod_{\substack{\alpha \neq 0 \\ \alpha \in F}}(x - \alpha)$$

or

$$x^q - 1 = \prod_{\alpha \neq 0}(x - \alpha).$$

Setting $x = 0$ we obtain

$$-1 = (-1)^{q-1}\prod_{\alpha \neq 0}\alpha$$

whence

$$\prod_{\alpha \neq 0}\alpha = (-1)^q.$$

Since either q is odd, or the characteristic is 2 and $-1 = +1$, we have

$$\prod_{\substack{\alpha \in F \\ \alpha \neq 0}}\alpha = -1,$$

which is a generalization of Wilson's theorem for R_p

$$(p-1)! \equiv -1 \pmod{p}.$$

We have seen that F is the splitting field of $x^q - x$ over R_p. Not only does F contain the roots of this polynomial—it consists entirely of the roots. There can be no smaller splitting field than F. So there is essentially no more than one field of degree n over R_p. For if F' also has p^n elements, then it too is a splitting field of the polynomial

$$x^{p^n} - x$$

and is hence isomorphic to F.

Conversely, if $q = p^n$ is given, we can construct a field of q elements—the splitting field F of $f(x) = x^{p^n} - x$ over R_p.

$$f'(x) = -1 \neq 0;$$

therefore there are no multiple roots. We may then write $f(x) = \prod_{i=1}^{p^n}(x - \alpha_i)$ where the α_i are distinct. F contains no other elements than the α_i since the α_i already constitute a field. In proof, consider any two of the α's, say α_1 and α_2. From $\alpha_1^{p^n} = \alpha_1$, $\alpha_2^{p^n} = \alpha_2$, we derive the rules:

Addition:[12] $\alpha_1 \pm \alpha_2 = \alpha_1^{p^n} \pm \alpha_2^{p^n} = (\alpha_1 \pm \alpha_2)^{p^n}$
Multiplication: $(\alpha_1\alpha_2)^{p^n} = \alpha_1\alpha_2$
Division: $(\alpha_1/\alpha_2)^{p^n} = \alpha_1/\alpha_2 \quad (\alpha_2 \neq 0)$
Sums, products, and quotients are again roots α.

We have proved

THEOREM 5.24 *To each power p^n of a prime p there is exactly one field (apart from isomorphism) with p^n elements. There are no other fields.*

In the further investigation of finite fields we shall require a number of group-theoretic lemmas.

LEMMA 5.25 *Let a, b be elements of a commutative group and denote their periods by α and β, respectively. Then there is an element c of the form $c = a^\nu b^\mu$ such that the period of c is the least common multiple of α and β.*

PROOF: Consider the factorization of α and β into primes (as on p. 31)

$$\alpha = q_1^{r_1} q_2^{r_2} \cdots q_m^{r_m},$$
$$\beta = q_1^{s_1} q_2^{s_2} \cdots q_m^{s_m} \quad (r_i, s_i \geq 0).$$

The least common multiple of α and β is

$$\gamma = q_1^{t_1} q_2^{t_2} \cdots q_m^{t_m}$$

where $t_i = \max(r_i, s_i)$. We choose ν and μ as follows:

$$\nu = q_1^{\delta_1} q_2^{\delta_2} \cdots q_m^{\delta_m},$$
$$\mu = q_1^{\epsilon_1} q_2^{\epsilon_2} \cdots q_m^{\epsilon_m},$$

[12] See the proposition on p. 57.

where if $t_i = r_i$ we take $\delta_i = 0$, $\epsilon_i = t_i$, and when $t_i \neq r_i$ we take $\delta_i = t_i$, $\epsilon_i = 0$. Denote the period of $c = a^\nu b^\mu$ by γ';

$$c^{\gamma'} = a^{\nu\gamma'} b^{\mu\gamma'}.$$

Now suppose $t_i = r_i$ so that $\delta_i = 0$, $\varepsilon_i = t_i$. Raising $c^{\gamma'}$ to the power

$$\frac{q_1^{t_1} q_2^{t_2} \cdots q_m^{t_m}}{q_i^{t_i}},$$

we then obtain

$$a^{\nu\gamma' \frac{q_1^{t_1} q_2^{t_2} \cdots q_m^{t_m}}{q_i^{t_i}}} b^{\mu\gamma' \frac{q_1^{t_1} q_2^{t_2} \cdots q_m^{t_m}}{q_i^{t_i}}} = 1.$$

But, by our assumption, μ has the factor $q_i^{\varepsilon_i} = q_i^{t_i}$. It follows that the exponent of b is certainly divisible by β. We conclude that

$$a^{\nu\gamma' \frac{q_1^{t_1} \cdots q_m^{t_m}}{q_i^{t_i}}} = 1$$

whence

$$\alpha \left| \frac{\nu\gamma' q_1^{t_1} \cdots q_m^{t_m}}{q_i^{t_i}} \right.$$

or, in particular,

$$q_i^{r_i} \left| \frac{\nu\gamma' q_i^{t_i} \cdots q_m^{t_m}}{q_i^{t_i}} \right. .$$

But $r_i = t_i$, $\delta_i = 0$. Therefore

$$q_0^{r_i} | \gamma'.$$

We immediately conclude that

$$\prod_{t_i = r_i} q_i^{t_i} | \gamma'.$$

On the other hand, if $t_i \neq r_i$ then $\delta_i = t_i = s_i$ and $\varepsilon_i = 0$. The above relation reduces to

$$b^{\mu\gamma' \frac{q_1^{t_1} q_2^{t_2} \cdots q_m^{t_m}}{q_i^{t_i}}} = 1$$

since the exponent of a is now clearly divisible by α. Consequently

$$q_i^{s_i} | \gamma'$$

where $s_i = t_i$. We have now proved that for the period γ' of c

$$q_i^{t_i} | \gamma'$$

for all i. So $\gamma | \gamma'$. However, obviously

$$(a^\nu b^\mu)^\gamma = 1$$

and therefore $\gamma' | \gamma$. This is possible only if $\gamma = \gamma'$. The proof is complete. □

LEMMA 5.26 *If G is an abelian (i.e., commutative) group and if the maximum period achieved by any element is $m > 0$, then*

$$x^m = 1$$

for all $x \in G$.

PROOF: Suppose a were an element with $a^m \neq 1$. Then the period α of a is such that $a \nmid m$. There is a $b \in G$ of period m. Hence by Lemma 5.25 we can construct an element c which has a greater period than m. □

EXAMPLE. If G is nonabelian this lemma is no longer true. Consider, for example, the case of the triangle group (Exercise 1, p. 1).

Consider any field F, finite or not. The set of all nonzero elements of F form an abelian group with respect to multiplication. Let G be a finite subgroup of the multiplicative group in F. Denote the order of G by N and denote by m the maximal period of its elements. By Lemma 5.26 the N elements of G all satisfy the equation $x^m = 1$. We have an equation of mth degree with at least N solutions and therefore $m \geq N$. But since m is the period of some element $m|N$, we have $m = N$; i.e., there is an element of G with period N. We have proved

THEOREM 5.27 *Every finite multiplicative subgroup of a field is cyclic.*

More precisely, we have proved for a group G of order N that the elements are the solutions of the equation $x^N = 1$. G consists of all the Nth roots of unity.

If now we restrict F to be finite, with say q elements, then the $q - 1$ nonzero elements form a multiplicative group. But this implies the

COROLLARY *The multiplicative group of a finite field is cyclic.*

There is an element of period $q - 1$ and it is therefore impossible to reduce the exponent in Fermat's theorem. In other words, in a finite field there is an element whose powers run through all the nonzero elements of the field. For the integers $(\text{mod } p)$ we have proved the existence of a primitive congruence root.

If the characteristic of F is p and if α is the element that generates F, then clearly $E = R_p(\alpha)$. Now q is a power of the characteristic (Lemma 5.23); $q = p^n$ where $n = (F/R_p)$. Therefore α is a root of an irreducible equation of nth degree. (We have thus proved the existence of irreducible polynomials of all degrees!) To obtain a field with p^n elements we have then to find an irreducible equation of nth degree and to adjoin a root to R_p.

EXAMPLE. Let us construct the field of 5^2 elements. The ground field is

$$R_5\colon 0, 1, 2, 3, 4.$$

The general polynomial of second degree is

$$[*] \qquad x^2 + ax + b.$$

(It is no restriction on the roots to assume the leading coefficients is 1.) Thus there are 25 equations of second degree. If the polynomial [*] is reducible at all it is the product of two linear factors. There are five possible factors, hence five polynomials with double roots, ten with distinct roots. The remaining ten polynomials are irreducible. We note immediately that $x^2 - 2$ and $x^2 - 3$ are irreducible. We are thereby provided with two ways of constructing the field of 5^2 elements. We must then be able to express $\sqrt{3}$ in terms of $\sqrt{2}$. Let θ denote a root of $x^2 - 2$. $R_5(\theta)$ consists of the elements $a + b\theta$ where $a, b \in R_5$, $\theta^2 = 2$. To represent $\sqrt{3}$ in $R_5(\theta)$ consider the equation $(a + b\theta)^2 = 3$, which is equivalent to

$$a^2 + 2b^2 + 2ab\theta = 3.$$

We can only have $a = 0$, $b = \pm 2$.

Let F be the field of nth degree over R_p. What are the automorphisms of F/R_p? One automorphism is $\sigma(\alpha) = \alpha^p$. By the proposition on page 57 we see that

$$\sigma(\alpha \pm \beta) = (\alpha \pm \beta)^p = \alpha^p \pm \beta^p = \sigma(\alpha) \pm \sigma(\beta),$$

and a similar result holds for multiplication. We have only to show that the correspondence is 1-1:

$$\sigma(\alpha) = \sigma(\beta) \Rightarrow \alpha^p = \beta^p \Rightarrow (\alpha - \beta)^p = 0 \Rightarrow \alpha = \beta.$$

Having one automorphism we may iterate until we get repetitions:

$$\sigma = \alpha^p, \quad \sigma^2 = \alpha^{p^2}, \quad \dots \quad , \quad \sigma^d = \alpha^{p^d}, \quad \dots \ .$$

If d is the period of σ, then $\sigma^d = I$, the identity automorphism. Thus d is the least integer for which

$$\alpha^{p^d} = \alpha$$

for all q elements α in F. The equation $x^{p^d} - x = 0$ must then have all q elements as solutions and therefore $p^d \geq q = p^n$. Hence $d \geq n$. But, on the other hand, $\alpha^{p^n} = \alpha$ for all α. Consequently, $\sigma^n(\alpha) = \alpha$ for all a, i.e., $\sigma^n = I$. The period of σ can be nothing other than n.

We have shown that the automorphisms

$$I, \sigma, \sigma^2, \dots, \sigma^{n-1}$$

are distinct. Since $(F/R_p) = n$ there are no others. We have proved the important

THEOREM 5.28 *The group of F/R_p is cyclic of order $n = (F/R_p)$. Hence F is normal over R_p and must be the splitting field of an irreducible separable polynomial.*

CHAPTER 6

Polynomials with Integral Coefficients

There is yet one question which has occurred repeatedly and has not been dealt with in any adequate way. This is the question as to whether any given polynomial in the rational field is irreducible. We cannot delay the answer to this question any longer—for otherwise we shall not be able to solve any special equations. Therefore we shall deviate from the main course of these lectures to discuss the topic of irreducibility.

Let R be any commutative ring with

(1) no divisors of zero,
(2) a unit element,
(3) unique factorization into primes;[1] i.e., every element of R is either zero, a unit, a prime, or a product of primes.

Let $R[x]$ denote the ring of all polynomials with coefficients in R. We write $a|f(x)$ if $a \in R$ is a divisor of all the coefficients of $f(x)$; i.e., if we have $f(x) = ag(x)$ where $g(x)$ has coefficients in the ring.

THEOREM 6.1 (Gauss) *If $p \in R$ is a prime and $f(x)$, $g(x) \in R[x]$, then*

$$\left.\begin{matrix} p\nmid f(x) \\ p\nmid g(x) \end{matrix}\right\} \Rightarrow p\nmid f(x)\cdot g(x).$$

PROOF: Write $f(x)$ and $g(x)$ in the form

$$f(x) = a_0 + a_1x + a_2x^2 + \cdots + a_jx^j + \cdots$$
$$g(x) = b_0 + b_1x + b_2x^2 + \cdots + b_kx^k + \cdots$$

where a_j and b_k are the first coefficients of the respective polynomials which are not divisible by p. Consider the coefficient c_{j+k} of x^{j+k} in $f(x)\cdot g(x)$

$$c_{j+k} = a_jb_k + a_0b_{j+k} + a_1b_{j+k-1} + \cdots + a_{j-1}b_{k+1} + a_{j+1}b_{k-1} + \cdots + a_{j+k}b_0.$$

Now $p\nmid a_jb_k$ but p divides all the other terms. It follows that $p\nmid c_{j+k}$. □

The greatest common divisor of the coefficients of a polynomial $f(x) \in R[x]$ is called the *content* of $f(x)$. If the content of $f(x)$ is 1, $f(x)$ is said to be *primitive*. Denoting the content of $f(x)$ by d, we write

$$f(x) = dg(x).$$

Thus any polynomial may be written as the product of a ring element with a primitive polynomial.

---

[1] R is therefore less special than a principal ideal ring.

LEMMA 6.2 *Any product of primitive polynomials is primitive.*

PROOF: If $f(x)$ and $g(x)$ are primitive, then $f(x) \cdot g(x)$ is primitive. For if $f(x) \cdot g(x)$ were divisible by any ring element, then it would be divisible by a prime p. The lemma follows directly from Theorem 6.1. □

If for $f(x), g(x) \in R[x]$ there is an $h(x) \in R[x]$ such that $f(x) = g(x) \cdot h(x)$, we say that $g(x)$ divides $f(x)$ in the *strong* sense, or simply $g(x)$ divides $f(x)$ and we write $g(x)|f(x)$. If there is a ring element a such that $g(x)$ divides $af(x)$, then $g(x)$ is said to divide $f(x)$ in the *weak* sense.

LEMMA 6.3 *If $g(x)$ is primitive and $g(x)$ divides $f(x)$ in the weak sense, then $g(x)$ divides $f(x)$ in the strong sense.*

PROOF: If $g(x)$ divides $f(x)$ in the weak sense, then there is an $a \in R$ and an $h(x) \in R[x]$ such that

$$af(x) = g(x) \cdot h(x).$$

We may put

$$h(x) = dh_0(x)$$

where d is the content of $h(x)$ and $h_0(x)$ is primitive. Similarly, we may write

$$f(x) = bf_0(x)$$

where $f_0(x)$ is primitive. Thus we obtain

$$abf_0(x) = dg(x) \cdot h_0(x)$$

where d is the content of the right side and ab is the content of the polynomial on the left. It follows that $d|ab$ and $ab|d$; d and ab are equal except for perhaps a unit factor. By including the proper unit factor in one of the polynomials, we may ensure $ab = d$. The fact that there are no divisors of zero permits the use of the cancellation law so that

$$f_0(x) = g(x) \cdot h_0(x)$$

from which follows

$$f(x) = bf_0(x) = g(x) \cdot bh_0(x).$$

This is the desired result. □

The ring R may be extended to the so-called quotient field F by the method of Lemma 5.22 (p. 75). We shall now refer to the elements of R as integers and those of F as rationals. The ring $R[x]$ of polynomials with integral coefficients is considered to be imbedded in the ring $F[x]$ of polynomials with rational coefficients.

If $f(x), g(x) \in R[x]$, $g(x)$ primitive, and if $g(x)$ divides $f(x)$ in $F[x]$; i.e., if there is an $h(x) \in F[x]$ such that

$$f(x) = g(x) \cdot h(x),$$

then $g(x)$ divides $f(x)$ in the weak sense and therefore in the strong sense.

PROOF: Let a denote the common denominator of the coefficients of $h(x)$. Clearly

$$af(x) = g(x) \cdot ah(x)$$

where a $h(x)$ is a polynomial with integral coefficients. It follows that $f(x)$ is divisible by $g(x)$ in the weak sense. □

We conclude

LEMMA 6.4 *If $f(x), g(x)$ are polynomials with integer coefficients, $g(x)$ primitive, then if $g(x)$ divides $f(x)$ in $F[x]$, the quotient polynomial has integer coefficients.*

If a fractional coefficient arises in the process of long division, we now can be sure that there is a remainder.

THEOREM 6.5 *If a polynomial with integer coefficients possesses factors in $F[x]$, it possesses factors in $R[x]$.*

PROOF: Assume $f(x) \in R[x]$ and $h(x), g(x) \in F[x]$ such that

$$f(x) = g(x) \cdot h(x).$$

We may obviously write

$$g(x) = \frac{a}{b} g_0(x), \quad h(x) = \frac{c}{d} h_0(x),$$

where $g_0(x), h_0(x)$ are primitive, $a, b, c, d \in R$. In that case,

$$f(x) = \frac{ac}{bd} g_0(x) h_0(x).$$

So $f(x)$ is weakly divisible by primitive polynomials, therefore strongly divisible. It follows that ac/bd is integral. □

In $R[x]$ there are two kinds of elements which may be called primes. These are the old primes in R, the constants, and the nonconstant, primitive irreducible polynomials.

Every polynomial possesses a unique factorization into prime constants and primitive irreducible polynomials. For the proof it is entirely sufficient to show that a prime which divides a product divides one of the factors.

THEOREM 6.6 *If P is a prime in $R[x]$ and $P | f(x) \cdot g(x)$, then either $P | f(x)$ or $P | g(x)$.*

PROOF:

(a) If $P \in R$ the proof is immediate by Theorem 6.1.
(b) Assume $P = P(x) \in R[x]$, $P(x)$ irreducible and primitive. But the theorem has already been proved for irreducible polynomials over a field. Therefore there exists an $h(x) \in F[x]$ such that, say,

$$g(x) = P(x)h(x).$$

By Lemma 6.4 it follows that $h(x) \in R[x]$ and the theorem is proved.

□

EXERCISE 1. What are the units in $R[x]$? Complete the proof that $R[x]$ is a unique factorization ring.

EXAMPLE. $R[x]$ satisfies the conditions of a unique factorization ring. Therefore we may adjoin a new variable to obtain a new ring in which unique factorization holds. This is the set of all polynomials in y whose coefficients are polynomials in x—the ring $R[x, y]$ of polynomials in two variables. Apparently, the ring of polynomials in n variables over a unique factorization domain is again a unique factorization domain. A field is a unique factorization domain (every element is either zero or a unit) so this remark applies to fields. We have proved even that the polynomials over a field form a principal ideal ring. This is not true for polynomials in more than one variable, however. For example, the set of all polynomials in two variables which are zero at the origin, i.e., have no constant term, definitely form an ideal—but it is not principal. For let $\mathfrak{K}$ be the set of all polynomials $f(x, y)$ which vanish for $x = 0$, $y = 0$. Clearly $x, y \in \mathfrak{K}$. But if $\mathfrak{K}$ consists of the multiples of a single element ϕ then $\phi | x$. Hence, either $\phi = x$ or $\phi = c$, a constant. Since y is also a multiple of ϕ we must have $\phi = c$. We cannot take $\phi = c$ since no nonzero constant is in the ideal.

6.1. Irreducibility

Let us consider a specific example, the polynomial $x^5 - x - 1$, in order to see what difficulties occur in proving irreducibility. Set

$$p(x) = x^5 - x - 1.$$

Does $p(x)$ have a linear factor? If so, it must have integer coefficients. Since the leading coefficient is 1 we may write

$$p(x) = (x - a)(x^4 + \cdots)$$

where a is an integer. Comparing terms, we see that $a|1$ so that $a = 1$ or $a = -1$. Neither is a root so the possibility of a linear factor is excluded.

The only remaining possibility is that $p(x)$ is the product of a quadratic and a cubic, say

$$p(x) = g(x) \cdot h(x)$$

where

$$g(x) = x^2 + ax + b, \quad h(x) = x^3 + cx^2 + dx + e,$$

a, b, c, d, e are integers. Let us see what possibilities there are for values of $g(x)$ for different values of x:

x	$p(x)$	$g(x)$
2	29	± 1 or ± 29
1	-1	± 1
0	-1	± 1
-1	-1	± 1
-2	-31	± 1 or ± 31

From the value at $x = 0$ we see that $b = \pm 1$. If $x = 1$, $g(x) = 1 + a \pm 1 = \pm 1$; therefore either $a = 1$ or -1 or -3. If $x = -1$, $g(x) = 1 - a + 1 = +1$; therefore $a = 1$ or -1 or $+3$. Hence $a = \pm 1$, and

$$g(x) = x^2 \pm x \pm 1.$$

Thus $|g(2)|$ cannot be 29, $|g(-2)|$ cannot be 31. $g(x)$ must take on either of the values ± 1 in five places. Therefore $g(x)$ must take on one value three times, which is impossible.

The same method is applicable to the general polynomial of the nth degree. We wish to determine whether a polynomial

$$f(x) = c_n x^n + c_{n-1} x^{n-1} + \cdots + c_0$$

has a factor of the rth degree, $0 < r < n$, say

$$g(x) = a_r x^r + a_{r-1} x^{r-1} + \cdots + a_0.$$

We must investigate at least $r + 1$ values of x since to determine a polynomial of rth degree we must fix $r + 1$ points. So we construct a table

x	$f(x)$	$g(x)$
x_0	f_0	$d_1^0, d_2^0, d_3^0, \ldots$
x_1	f_1	$d_1^1, d_2^1, d_3^1, \ldots$
$\vdots$	$\vdots$	$\vdots$
x_r	f_r	$d_1^r, d_2^r, d_3^r, \ldots$

where the d_j^k are the divisors of f_k. It is clearly to our advantage to choose values of x for which $f(x)$ is prime and large. The method is now to interpolate polynomials through the possible values of $g(x)$, e.g.,

$$g(x_0) = d_1^0, \quad g(x_1) = d_1^1, \quad \ldots \quad , \quad g(x^r) = d_1^r.$$

If an interpolation does not lead to a polynomial with integral coefficients, we can reject it at once. Otherwise we must test by long division into $f(x)$ or by expanding our table. The method must be repeated for all possible combinations of the d's to be a sufficient proof of irreducibility.

This approach is obviously the last resort of the desperate. We shall soon discuss certain sufficient conditions for irreducibility which are often of great use.

EXAMPLE. Let us determine the values of a for which

$$f(x) = x^5 - ax - 1$$

is irreducible. If $f(x)$ has linear factors it must have either $+1$ or -1 as a root

$$\begin{aligned} 1 - a - 1 &= 0; \quad & a &= 0 \\ -1 + a - 1 &= 0; \quad & a &= -2. \end{aligned}$$

If $f(x)$ has a quadratic factor we may write

$$f(x) = x^5 - ax - 1 = (x^2 + bx + c)(x^3 + dx^2 + ex + f).$$

Equating the coefficients of the terms of equal degree we obtain the relations

$$[1] \qquad b + d = 0,$$

$$[2] \qquad e + bd + c = 0,$$

$$[3] \qquad f + be + cd = 0,$$

$$[4] \qquad bf + ce = -a,$$

$$[5] \qquad cf = -1.$$

Equation [5] yields

$$c = -f = \pm 1.$$

Using $b = -d$ (from [1]) in [3] we obtain

$$d(c - e) = -f = \pm 1,$$

whence $d = +1$ and $(c - e) = \pm 1$. From [2]

$$e + c = 1, \quad e - c = \pm 1,$$

whence either $c = 0$ or $c = 1$. The first case is impossible if we are to satisfy [5]. So we obtain at once:

$$c = 1, \quad e = 0, \quad f = -1, \quad d = 1, \quad b = -1.$$

From (4) we have $a = -1$. There is then only one possibility for a quadratic factor:

$$x^5 + x - 1 = (x^2 - x + 1)(x^3 + x^2 - 1).$$

There are only three reducible cases: $a = 0$, $a = 2$, $a = -1$. In the first two cases there cannot be a quadratic factor so $f(x)$ factors into a linear and a fourth degree factor.

Consider another example. For what values of a is the polynomial

$$f(x) = x^5 - x - a$$

reducible? There are obviously an infinite number of possibilities for linear factors for we need only take $a = b^5 - b$ is any integer.

EXERCISE 2. For how many values of a does the polynomial

$$f(x) = x^5 - x - a$$

have a quadratic factor?

Hint: This problem leads to a diophantine equation which has only a few solutions.

THEOREM 6.7 (Eisenstein) *If for the polynomial*

$$f(x) = a_n x^n + a_{n-1}x^{n-1} + \cdots + a_0$$

there is a prime p such that

(1) $p \nmid a_n$,
(2) $p | a_i \quad \text{for } i = 0, 1, 2, \ldots, n-1$,
(3) $p^2 \nmid a_0$,

then $f(x)$ is irreducible.

PROOF: Assume $f(x) = \phi(x) \cdot \psi(x)$ is a factorization of $f(x)$ into polynomials of positive degree. Since the degrees of ϕ and ψ are less than n, we may write

$$\phi(x) = b_0 + b_1 x + \cdots + b_{n-1}x^{n-1},$$
$$\psi(x) = c_0 + c_1 x + \cdots + c_{n-1}x^{n-1},$$

where some of the coefficients may be zero.

Now $a_0 = b_0 c_0$ and $p | a_0$, therefore $p | b_0$ say. Since $p^2 \nmid a_0$ then $p \nmid c_0$.

Continue in the same manner: $a_1 = c_0 b_1 + c_1 b_0$. Since $p | c_1 b_0$ and $p | a_1$ we must have $p | c_0 b_1$. But $p \nmid c_0$ and therefore $p | b_1$. Clearly we can prove that all the b's are divisible by p. Consequently, $p | \phi(x)$, which implies $p | f(x)$ in contradiction to the assumption that $p \nmid a_n$. □

6.2. Primitive Roots of Unity

Let us examine the equation

$$x^p - 1 = 0$$

over the field R of rational numbers. This has the obvious root 1 and so we may write

$$x^p - 1 = (x-1)f(x) = (x-1)(x^{p-1} + x^{p-2} + \cdots + x + 1).$$

The irreducibility of $f(x)$ may be proved by Eisenstein's criterion. Write $f(x)$ in the form

$$f(x) = \frac{x^p - 1}{x - 1}.$$

Now the question of the reducibility of $f(x)$ is clearly equivalent to that of $f(x+1)$ so it is sufficient to investigate

$$f(x+1) = \frac{(x+1)^p - 1}{x}.$$

By an application of the binomial theorem this yields

$$f(x+1) = x^{p-1} + \binom{p}{1}x^{p-2} + \cdots + \binom{p}{2}x + p.$$

Since the binomial coefficients are divisible by p (cf. p. 57) and the last term p is not divisible by p^2, our assertion is proved.

Let us now attempt a similar proof for the exponent p^n, $n > 1$,

$$x^{p^n} - 1 = 0.$$

This equation is clearly satisfied by the $(p^{n-1})^{\text{th}}$ roots of unity so that

$$x^{p^n} - 1 = (x^{p^{n-1}} - 1)\phi(x).$$

We shall prove that

$$\phi(x) = \frac{x^{p^n} - 1}{x^{p^{n-1}} - 1}$$

is irreducible. Put

$$\phi(x) = f(x^{p^{n-1}}) = f(y)$$

where

$$y = x^{p^{n-1}}, \quad f(y) = \frac{y^p - 1}{y - 1}.$$

We have

$$f(y+1) = y^{p-1} + (\text{terms divisible by } p) + p.$$

Now

$$\begin{aligned}
\phi(x+1) &= f\big([x+1]^{p^{n-1}}\big) \\
&= f\big(\big[x^{p^{n-1}} + (\text{terms divisible by } p)\big] + 1\big) \\
&= \big[x^{p^{n-1}} + (\text{terms divisible by } p)\big]^{p-1} + (\text{terms divisible by } p) + p \\
&= x^{(p-1)p^{n-1}} + (\text{terms divisible by } p) + p.
\end{aligned}$$

The proof is immediate by Eisenstein's criterion.

The proof shows that

$$x^{p^n} - 1 = (x^{p^{n-1}} - 1)\phi(x)$$

where $\phi(x)$ is irreducible. The formula gives the factorization recursively. So for all intents and purposes the factorization is complete. We have shown that the $(p^n)^{\text{th}}$ roots of unity include all $(p^m)^{\text{th}}$ roots of unity for $m \leq n$. Clearly $\phi(x)$ contains all the proper $(p^n)^{\text{th}}$ roots—those which are not of lower order. The roots of $\phi(x)$ are called the *primitive* roots of unity.

By Theorem 5.27 (p. 79) it is clear that the multiplicative group of the $(p^n)^{\text{th}}$ roots of unity is cyclic. The generators of the group are the primitive roots, for no root of lower order can possibly generate a primitive root and clearly no exponent less than p^n can be a period for a primitive root. The splitting field of $x^{p^n} - 1$ is then just the splitting field of $\phi(x)$ and is obtained simply by adjoining a primitive root ε, $\phi(\varepsilon) = 0$, to the ground field R. The degree of the splitting field is that of the irreducible polynomial $\phi(x)$;

$$(R(\varepsilon)/R) = (p-1)p^{n-1}.$$

These methods may easily be extended to any m^{th} roots of unity. The roots of

$$x^m - 1 \tag{1}$$

form a cyclic group of order m. If ε is a generator, then the roots are $1, \varepsilon, \varepsilon^2, \ldots, \varepsilon^{m-1}$. When m is not prime, there are always roots of lower order—if d is a divisor of m, then $\varepsilon^{m/d}$ is a lower-order root of period d. The primitive roots are clearly those ε^ν where $(\nu, m) = 1$ (read: "ν is prime to m" for "$(\nu, m) = 1$"). The polynomial for the primitive m^{th} roots of unity is therefore

$$\Phi_m(x) = \prod_{\substack{(\nu,m)=1 \\ 0<\nu<m-1}} (x - \varepsilon^\nu).$$

The factorization of [1] is then given by

[2] $$x^m - 1 = \prod_{d|m} \Phi_d(x).$$

It is very easy to compute a table of the Φ's:

$$\begin{aligned}
\Phi_1(x) &= x - 1, \\
\Phi_2(x) &= x + 1, \\
\Phi_3(x) &= x^2 + x + 1, \\
\Phi_4(x) &= x^2 + 1, \\
\Phi_5(x) &= x^4 + x^3 + x^2 + x + 1, \\
\Phi_6(x) &= x^2 - x + 1, \\
&\vdots
\end{aligned}$$

EXERCISE 3.

(a) Prove that $\Phi_{2m}(x) = \Phi_m(-x)$ for m odd.

(b) Prove that if $p \nmid m$, where p is prime, then $\Phi_{pm}(x) = \Phi_m(x^p)/\Phi_m(x)$. Discuss the case when $p|m$.

The polynomials Φ_m are called the *cyclotomic* (circle-dividing) polynomials. The field $R(\varepsilon)$, where ε is a root of a cyclotomic polynomial, is sometimes called a cyclotomic field.

CHAPTER 7

The Theory of Equations

This discussion has reached a point of development where the theory may be used fruitfully to solve a number of important problems. We begin with the problem of

7.1. Ruler and Compass Constructions

Among the problems passed down by the ancient Greeks there are the two familiar ones of the trisection of the angle and the duplication of the cube. The problem of dividing an angle into three equal parts, in particular, has frequently been "solved" by numerous well-meaning but guileless souls who have not troubled to discover what exactly is the problem. The ancients already had a simple means of trisecting angles:

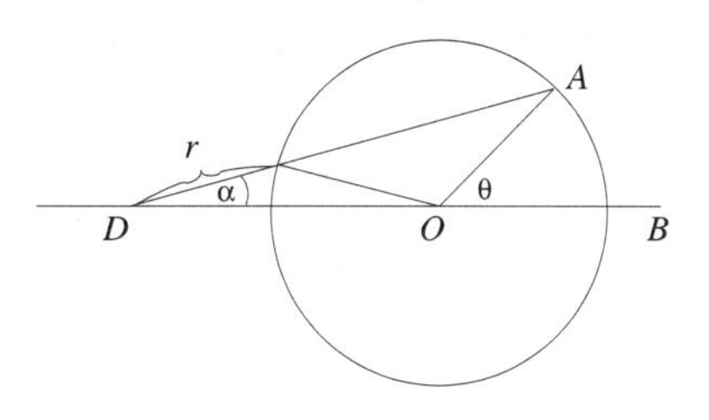

In the figure, let $\theta = \measuredangle AOB$ be any angle. Draw a circle about O of arbitrary radius r. Mark off the length r on a straight edge. Place the ruler with its edge on A, one mark on the line OB and the other on the circle (see figure). It is easily verified that $\alpha = \measuredangle ADO = \theta/3$. But this is not a solution of the problem.

In a geometrical construction we are given various initial data, points and line segments, and we seek to determine other configurations by means of a finite number of admissible operations on the given information. What operations are admitted:

(a) marking of arbitrary points,
(b) drawing a line between two points (the only allowable use of the ruler),
(c) drawing a circle with a given radius and a given point as center,
(d) determining a point as the intersection of two lines or two circles or a line and a circle.

The given data will include various lengths $x_1, x_2, \ldots, x_n$. Using one of the data as a unit, we construct a Cartesian coordinate system. To do this we need only pick an arbitrary pair of points, draw the line through them, and erect a perpendicular anywhere on the line. Clearly, we can construct any point with integer coordinates. It is a simple matter to construct any point with rational coordinates since a segment can be divided into any number of equal parts. We shall restrict the choice of arbitrary points to points with rational coordinates because these are constructible and since any construction which cannot be performed with this restricted choice of points is certainly impossible for a completely arbitrary choice.

Given any lengths x, y we can easily construct the sum $x + y$ and the difference $x - y$, and (see figure) the product and quotient, xy and x/y. It follows that we can construct any element of the field $R(x, y)$ generated by these elements. In general, if $x_1, x_2, \ldots, x_n$ are the given data, we can certainly construct any element in the field $R(x_1, x_2, \ldots, x_n)$. We may extend the set of constructible elements further by considering the intersections of circles with straight lines or with circles. The intersection of two straight lines gives nothing new. The coordinates of a point on a straight line satisfy a linear relation. Hence the determination of a point as the intersection of two straight lines involves the solution of a pair of linear equations and does not take us out of the ground field. The problem of two intersecting circles can be reduced to the intersection of a straight line on a circle and this will usually necessitate going outside the field. The determination of a point as the intersection of a circle with a straight line involves the solution of a quadratic equation. Should this be irreducible, we append the solution to the ground field and thereby obtain an extension of degree 2.

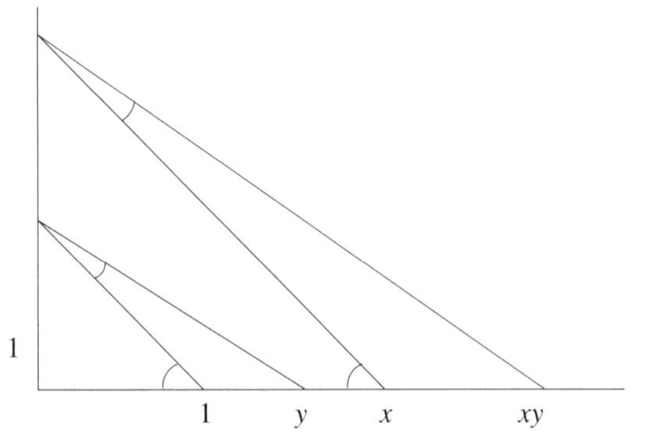

Thus, in any construction, we begin with certain accessible elements, the elements of the ground field $R(x_1, x_2, \ldots, x_n)$, and at each step we obtain a new field of accessible elements where, if say F_n is the field at the nth step, we have

$$(F_{n+1}/F_n) \geq 2.$$

Consequently, if a construction can be performed in n steps, the degree of F_n over the ground field must be a power of 2,

$$(F_n/F) = 2^\nu \quad (\nu \leq n).$$

Suppose the solution requires a segment of length α, e.g., α is the chord on the unit circle subtended by the angle $\theta/3$. The length α must be an element of F_n and therefore the field $F(\alpha)$ is intermediate between F and F_n,

$$F_n \supset F(\alpha) \supset F.$$

The degree of $F(\alpha)$ is a divisor of the degree of F_n and therefore must also be a power of 2,

$$(F(\alpha)/F) = 2^\mu \quad (\mu \leq \nu).$$

This is then a necessary condition that a given construction be possible. The construction must not involve the determination of any length which leads to an extension of a degree other than a power of two.

This result may immediately be applied to the famous problem of the Delphian oracle, namely, the construction of a cube twice the size of a given cube. If we accept the side length of the given cube as the unit, this problem is equivalent to solving the irreducible equation

$$x^3 = 2.$$

Since $(R(\sqrt[3]{2})/R) = 3$, the construction cannot be performed.

We have found a necessary condition that a length α be constructible, namely, that

$$(F(\alpha)/F) = 2^\nu.$$

Thus, in many cases we can prove the impossibility of a construction. It is natural to ask for a condition that is both necessary and sufficient. Since any constructible length can be derived from the data of the problem by the rational operations and extractions of square roots, this condition is that it be possible to find a field E containing α for which there is a chain of fields with E at the top and F at the bottom,

$$F = F_1 \subset F_2 \subset \cdots \subset F_n = E,$$

and $(F_{\nu+1}/F_\nu) \le 2$.

Let us examine the problem of constructing the m-gon, the regular polygon of m sides. Consider the factorization of m into primes,

$$m = p_1^{\nu_1} p_2^{\nu_2} \cdots p_r^{\nu_r}.$$

If the m-gon can be constructed, then plainly we can construct any d-gon where d is a divisor of m. In particular, we can construct the polygons of $p_1^{\nu_1}, p_2^{\nu_2}, \ldots, p_r^{\nu_r}$ sides. Conversely, if it is possible to construct these $p_i^{\nu_i}$-gons $(i = 1, 2, \ldots, r)$, then we can construct the m-gon. For the numbers

$$m/p_1^{\nu_1}, m/p_2^{\nu_2}, \ldots, m/p_r^{\nu_r}$$

are relatively prime and therefore the diophantine equation

$$mx_1/p_1^{\nu_1} + mx_2/p_2^{\nu_2} + \cdots + mx_r/p_r^{\nu_r} = 1$$

has a solution in integers $x_1, x_2, \ldots, x_r$. Dividing by m we obtain $\sum(x_i/p_i^{\nu_i}) = 1/m$. Hence an mth part of a circle consists of a sum of $(p_i^{\nu_i})^{\text{th}}$ parts. We need therefore consider only powers of primes.

EXAMPLE. The problem of constructing a 15-gon reduces by these considerations to the problem of constructing an equilateral triangle and a regular pentagon. We must find integers x, y with

$$5x + 3y = 1.$$

The numbers $x = -1$, $y = 2$ work, $-\frac{1}{3} + \frac{2}{5} = \frac{1}{15}$. To construct the 15th part of a circle we first construct an angle of 144° and then subtract 120°.

REMARK. We shall say an imaginary number $a+ib$ is *constructible* if the real and imaginary parts separately are constructible. The introduction of imaginary elements does not affect our theory. The sum, difference, product, and quotient of two constructible imaginaries are constructible. Furthermore, the square root of a constructible imaginary is constructible,

$$\sqrt{a+ib} = \sqrt{\frac{a+\sqrt{a^2+b^2}}{2}} + i\sqrt{\frac{-a+\sqrt{a^2+b^2}}{2}}.$$

The problem of constructing an m-gon is equivalent to the problem of constructing the length $\cos\frac{2\pi}{m}$. If this length is constructible, so is $\sin\frac{2\pi}{m}$ and also

$$\varepsilon = \cos\frac{2\pi}{m} + i\sin\frac{2\pi}{m}.$$

ε is a root of the equation $\varepsilon^m = 1$. If m is a power of a prime, $m = p^\nu$, then, by the results starting on page 87, ε is a root of the irreducible polynomial

$$\phi(x) = \frac{x^{p^\nu} - 1}{x^{p^{\nu-1}} - 1}$$

which is of degree $p^{\nu-1}(p-1)$. But we can construct ε only if the degree of the extension field $R(\varepsilon)$ is a power of 2, i.e., only if

$$p^{\nu-1}(p-1) = 2^\mu.$$

We can have only $p = 2$ or $\nu = 1$. Except for 2, no power of a prime higher than the first can be admitted. In particular, the polygon of 9 sides cannot be constructed and it is therefore impossible to trisect the angle of 120°. As a side result, we have shown that the trisection problem cannot have a general solution.

For $\nu = 1$ we have $p = 2^\mu + 1$. We are interested in all primes of this form. If μ has an odd divisor, $\mu = \lambda(2n+1)$, then

$$2^\mu + 1 = 2^{\lambda(2n+1)} + 1 = (2^\lambda + 1)(\cdots)$$

is not a prime. The only primes which are allowable must therefore have the form

$$p = 2^{2^k} + 1.$$

The numbers of this form are prime for values of k up to 4. These are the primes $3, 5, 17, 257, 65537$. Fermat's famous conjecture is that these numbers are prime for all values of k. Actually, this breaks down at 5; $2^{32} + 1$ is divisible by 641. It is not known whether there are an infinite number of primes of this form. In any case, the only constructible m-gons are those for which m is a product of powers of 2 with Fermat primes, none of the latter appearing to a power higher than the first. We have really only shown that no other polygons are constructible. In order to show that the construction of these polygons is possible, we shall need more refined tools.

7.2. Solution of Equations by Radicals

We consider a more general problem than that of ruler and compass construction. When can an equation be solved completely in terms of rational operations and root extractions? In order to answer this question, we must develop some preliminary results.

Let E/F be normal, and G the group of automorphisms of E/F. We shall employ the following:

Notation. If $\alpha \in E, \sigma \in G$, we set

$$\sigma(\alpha) = \alpha^{\sigma}.$$

In general, if $a_1, a_2, \dots, a_n$ are integers, we shall write

$$\sigma_1(\alpha^{a_1})\sigma_2(\alpha^{a_2})\cdots\sigma_n(\alpha^{a_n}) = \alpha^{a_1\sigma_1+a_2\sigma_2+\cdots+a_n\sigma_n}.$$

For example, by the expression $\alpha^{I+\sigma_1-3\sigma_2}$ we mean

$$\frac{I(\alpha)\cdot\sigma_1(\alpha)}{\sigma_2(\alpha^3)} = \frac{\alpha\cdot\sigma_1(\alpha)}{[\sigma_2(\alpha)]^3}.$$

Furthermore, we have

$$\begin{aligned}[\alpha^{a_1\sigma_1+a_2\sigma_2+\cdots+a_n\sigma_n}]^{\tau} &= \tau[\sigma_1(\alpha^{a_1})\sigma_2(\alpha^{a_2})\cdots\sigma_n(\alpha^{a_n})] \\ &= \alpha^{a_1\tau\sigma_1+a_2\tau\sigma_2+\cdots+a_n\tau\sigma_n},\end{aligned}$$

and this we set equal to

$$\alpha^{\tau(a_1\sigma_1+a_2\sigma_2+\cdots+a_n\sigma_n)}.$$

Hence, in general, we define

$$[\alpha^{\Sigma a_\nu\sigma_\nu}]^{\Sigma b_\mu\tau_\mu} = \prod_\mu [(\alpha^{\Sigma a_\nu\sigma_\nu})^{\tau_\mu}]^{b_\mu} = \prod_\mu \alpha^{b_\mu\tau_\mu\Sigma a_\nu\sigma_\nu} = \alpha^{\Sigma b_\mu\tau_\mu\Sigma a_\nu\sigma_\nu}.$$

Now, if the group of E/F is $G = \{\sigma_1, \sigma_2, \dots, \sigma_n\}$, we define the *norm* $N\alpha$ for $\alpha \in E$ to be the product of the images of α under G,

$$N\alpha = \sigma_1(\alpha)\sigma_2(\alpha)\cdots\sigma_n(\alpha) = \alpha^{\sigma_1+\sigma_2+\cdots+\sigma_n}.$$

We must have $N\alpha \in F$ since

$$(N\alpha)^{\tau} = \alpha^{\sigma\tau_1+\sigma\tau_2+\cdots+\sigma\tau_n}$$

and the $\tau\sigma_i$ are simply the elements of G permuted. If G is cyclic of order n,

$$G = \{I, \sigma, \sigma^2, \dots, \sigma^{n-1}\},$$

we have (using $N(\alpha\beta) = N\alpha\cdot N\beta$)

$$N\alpha^{I-\sigma} = (N\alpha)^{I-\sigma} = \alpha^{(I-\sigma)(1+\sigma+\sigma^2+\cdots+\sigma^{n-1})} = \alpha^{I-\sigma^n} = \alpha^0 = 1.$$

What are all the elements α for which $N\alpha = 1$? Clearly, if $\alpha = \beta^{I-\sigma}$ for any $\beta \in E$, we have

$$N\alpha = N\beta^{1-\sigma} = 1.$$

That these are all such elements is proved in a remarkable theorem of Kummer's.

THEOREM 90[1] *If the group G of E/F is cyclic with generator σ, then the elements α in E with the norm* 1 *are precisely those which can be written in the form*

$$\alpha = \beta^{I-\sigma} \quad (\beta \in E,\ \beta \neq 0).$$

[1] So-called because of its appearance under that number in Hilbert's *Zahlbericht*.

PROOF: The norm of any element $\beta^{I-\sigma}$, $\beta \neq 0$, is certainly 1 for

$$\beta^{I-\sigma} = \frac{\beta}{\sigma(\beta)},$$

whence

$$N(\beta^{I-\sigma}) = N\beta / N\sigma(\beta) = 1.$$

Conversely, if $N\alpha = 1$ there is a $\beta \in E$ with $\alpha = \beta^{I-\sigma}$. Put

$$\beta = 1 + \alpha + \alpha^{I+\sigma} + \alpha^{I+\sigma+\sigma^2} + \cdots + \alpha^{I+\sigma+\cdots+\sigma^{n-2}}.$$

We have

$$\alpha\beta^{\sigma} = \alpha + \alpha^{I+\sigma} + \alpha^{I+\sigma+\sigma^2} + \cdots + \alpha^{I+\sigma+\cdots+\sigma^{n-1}} = \beta$$

since $N\alpha = \alpha^{I+\sigma+\cdots+\sigma^{n-1}} = 1$. Thus β satisfies the relation $\alpha\beta^{\sigma} = \beta$, and if $\beta \neq 0$ it follows that $\alpha = \beta^{I-\sigma}$. But β might very well be zero. This difficulty is disposed of by including extra factors. Let θ be arbitrary in E and set

$$\beta = \theta + \theta^{\sigma}\alpha + \theta^{\sigma^2}\alpha^{I+\sigma} + \cdots + \theta^{\sigma^{n-1}}\alpha^{I+\sigma+\cdots+\sigma^{n-2}}.$$

Consider the expression

$$\alpha\beta^{\sigma} = \theta^{\sigma}\alpha + \theta^{\sigma^2}\alpha^{I+\sigma} + \cdots + \theta^{\sigma^n}\alpha^{I+\sigma+\cdots+\sigma^{n-1}}.$$

From $\alpha^{I+\sigma+\cdots+\sigma^{n-1}} = N\alpha = 1$, $\theta^{\sigma^n} = 0$, we have $\alpha\beta^{\sigma} = \beta$. If it is possible to find a θ for which $\beta \neq 0$, we have proved our theorem.

Suppose the contrary, that $\beta = 0$ for every value of θ; i.e., that

$$\theta + \theta^{\sigma}\alpha + \theta^{\sigma^2}\alpha^{I+\sigma} + \cdots + \theta^{\sigma^{n-1}}\alpha^{I+\sigma+\cdots+\sigma^{n-2}} = 0$$

for all θ. This is a linear relation in $\theta, \sigma(\theta), \sigma^2(\theta), \ldots, \sigma^{n-1}(\theta)$, but no such relation can exist (Lemma 5.11, p. 61). The theorem is proved. □

EXAMPLE. The field $R(i)$ where i is a root of the equation $x^2 + 1 = 0$ over the rational field R has precisely two automorphisms, the identity and σ, where $\sigma(a + bi) = a - bi$. We have

$$N(3 + 4i)/5 = (3 + 4i)/5 \cdot (3 - 4i)/5 = 1.$$

Consequently, there is a β with

$$(3 + 4i)/5 = \beta^{I-\sigma}.$$

The simplest possibility is $\theta = 1$, $\beta = 1 + \alpha$:

$$(1 + \alpha)^{I-\sigma} = \left(\frac{8 + 4i}{5}\right)^{I-\sigma} = (8 + 4i)/(8 - 4i) = (3 + 4i)/5.$$

It is extremely useful to know the set of elements α for which $N\alpha = 1$. We would like to know what happens when the group is not cyclic. Unfortunately, though many attempts have been made to generalize the theorem to arbitrary groups, no answer to the problem has been provided. It is even doubtful that a general answer can be found.

EXERCISE 1.

(a) Suppose E/F normal, $(E/F) = n$. Consider the n^2 equations

$$x_{\sigma\tau} = x_\sigma x_\tau^\sigma$$

in the n unknowns x_σ. These have the nontrivial solution $x_\sigma = \beta^{I-\sigma}$, $\beta \in E$, $\beta \neq 0$. Show that this solution is unique.

(b) Why does this problem reduce to Theorem 7.2 in the cyclic case?

The usefulness of Theorem 7.2 is apparent in the following applications. Assume

(1) E/F is normal and cyclic of degree n with the group $I, \sigma, \ldots, \sigma^{n-1}$.

(2) F contains n distinct n^{th} roots of unity $1, \varepsilon, \varepsilon^2, \ldots, \varepsilon^{n-1}$.

The field E must then be the splitting field of an irreducible equation of the form

$$x^n - b = 0$$

where $b \in F$.

PROOF: Using the fact that ε is in the fixed field, we have

$$N\varepsilon^{-1} = \varepsilon^{-(I+\sigma+\cdots+\sigma^{n-1})} = \varepsilon^{-n} = 1.$$

It follows that there is a $b \in E$, $\beta \neq 0$, such that $\varepsilon^{-1} = \beta^{I-\sigma}$. Hence $\varepsilon^{-1} = \beta/\sigma(\beta)$ or $\sigma(\beta) = \varepsilon\beta$. We may therefore write

$$\begin{aligned} I(\beta) &= \beta, \\ \sigma(\beta) &= \varepsilon\beta, \\ \sigma^2(\beta) &= \varepsilon^2\beta, \\ &\vdots \\ \sigma^{n-1}(\beta) &= \varepsilon^{n-1}\beta, \end{aligned}$$

and these elements are distinct. Consequently, β satisfies an equation of degree n over F (Theorem 5.16, p. 66) and therefore $E = F(\beta)$. Furthermore, $\sigma(\beta^n) = [\sigma(\beta)]^n = \varepsilon^n\beta^n = \beta^n$; i.e., β^n is fixed in the automorphisms of E/F and therefore $\beta^n \in F$. Consequently, β is the root of an equation

$$x^n = b \quad (b \in F).$$

The field E cyclic of degree n over F may be obtained simply by adjoining one radical, $\beta = \sqrt[n]{b}$. Since β is the root of an irreducible equation of nth degree, we conclude that

$$x^n - b = (x - \beta)(x - \varepsilon\beta)\cdots(x - \varepsilon^{n-1}\beta). \qquad \square$$

EXAMPLE. Consider the field $R(\sqrt{11})$ of degree 2 over R. The primitive square root of 1 is in R, $\varepsilon = -1$. $N(-1) = (-1)^2 = 1$. Thus $N\varepsilon^{-1} = 1$. We can therefore write $-1 = \beta^{1-\sigma}$, $\beta \neq 0$, where β has the form

$$\beta = \theta + \theta^\sigma\varepsilon = \theta - \theta^\sigma.$$

$\theta = 1$ is not acceptable. Consequently, $\theta = \sqrt{11}$ must work, and in fact yields $\beta = 2\sqrt{11}$.

The result we have just obtained is constructive in that we may single out the element which generates the field. For β we take

$$\beta = \theta + \varepsilon^{-1}\theta^{\sigma} + \varepsilon^{-2}\theta^{\sigma^2} + \cdots + \varepsilon^{-(n-1)}\theta^{\sigma^{n-1}}$$

since $\sigma(\varepsilon) = \varepsilon$. Now β cannot be zero for all n basis elements of E over F. Hence we need to try at most n values of θ to determine β. Having found β, we obtain β^n and this must be an element of F.

Let us attack the converse problem. If F is a field containing a primitive n^{th} root of unity, what is the splitting field of the equation

$$\psi(x) = x^n - b = 0?$$

Let β be any root of $\psi(x)$ and form the field $E = F(\beta)$. Since the distinct elements $\beta, \varepsilon\beta, \ldots, \varepsilon^{n-1}\beta$ are all roots of $\psi(x)$, we conclude that E is the splitting field. Now two possible cases may arise:

Case 1. $\psi(x)$ is reducible.

In that event let $\phi(x)$ be the irreducible factor over F, with $\phi(\beta) = 0$. $\phi(x)$ will be responsible for a number of the roots

$$\phi(x) = \prod_{\text{some } \nu} (x - \varepsilon^{\nu}\beta).$$

The constant term in $\phi(x)$, an element of F, must be of the form $\pm\varepsilon^{\mu}\beta^r$ where $r = \partial[\phi(x)]$. Consequently, $\beta^r = c \in F$; i.e., β satisfies the equation $x^r - c = 0$ of rth degree over F. Since $\phi(x)$ is irreducible of degree r, it follows (Lemma 4.5, p. 40) that

$$\phi(x) = x^r - c.$$

Now $E = F(\beta) = F(\varepsilon\beta) = \cdots = F(\varepsilon^{n-1}\beta)$ since $\varepsilon \in F$. It is clear that every $\varepsilon^{\nu}\beta$ satisfies an irreducible equation of rth degree,

$$x^r - (\varepsilon^{\nu}\beta)^r = x^r - \varepsilon^{\nu r}c = 0.$$

We conclude that $\psi(x)$ factors into polynomials all of the same degree r and consequently that $r|n$. We have $\beta^r = c \in F$. Putting $n = rs$, we obtain

$$\beta^n = \beta^{rs} = c^s = b.$$

Hence the reducible case occurs only if b is a power of c^s, $s|n$. Conversely, it is evident if $b = c^s$, $s|n$, that the polynomial $\psi(x)$ may be factored into polynomials of equal degree $r = n/s$.

The reducible case is included in the irreducible case since F must contain a primitive r^{th} root of unity if it contains a primitive n^{th} root.

A word of caution. These criteria can only be applied when F contains a primitive n^{th} root of unity. Consider, for example, the polynomial $x^4 + 4$ over the rational field R. The number -4 cannot be expressed as the square or fourth power of any element in R, yet we have

$$x^4 + 4 = (x^2 - 2x + 2)(x^2 + 2x + 2).$$

Nevertheless, a germ of the result remains even when the roots of unity are not contained in the ground field. We digress to discuss the one case in which we may dispense with the roots of unity.

Let

$$\psi(x) = x^p - b, \quad p \text{ prime,}$$

and suppose that p is not the characteristic of F. Unless b is a pth power of some element in F, $\psi(x)$ is irreducible.

PROOF: Let E be the splitting field of $\psi(x)$. The roots of $\psi(x)$ are certainly distinct since $\psi'(x) = 0$ only if $x = 0$ and zero is not a root. Let β be any one of the roots. The roots of $\psi(x)$ are then

$$\beta,\ \varepsilon\beta,\ \varepsilon^2\beta,\ \dots,\ \varepsilon^{p-1}\beta,$$

where ε is a primitive p^{th} root of unity. Hence the splitting field of $x^p - b$ must contain the pth roots of unity.

Now, if $\psi(x)$ could be factored in F, we could find some irreducible factor of lower degree, say

$$\phi(x) = \prod_{\text{some } \nu} (x - \varepsilon^\nu \beta).$$

The constant term in $\phi(x)$, an element of F, is of the form $\pm\varepsilon^\mu\beta^r$ where $r = \partial[\phi(x)] < p$. We have

$$\varepsilon^\mu\beta^r = c \in F, \quad \beta^p = b \in F.$$

Now $r < p$, p prime $\Rightarrow (r, p) = 1$. It follows that the equation

$$rx + py = 1$$

has a solution in integers x, y,

$$c^x b^y = (\varepsilon^\mu\beta^r)^x(\beta^p)^y = \varepsilon^{\mu x}\beta \in F.$$

Clearly, then, if $\psi(x)$ is reducible it has a root in the ground field and b is after all a pth power in F. □

Case 2. $\psi(x)$ is irreducible.

We are concerned with the splitting field E of an irreducible equation of the form

$$\psi(x) = x^n - b = 0$$

where the ground field F is assumed to contain a primitive n^{th} root of unity. There are n possible automorphisms of F, these being given by the transformations

$$\sigma_\nu(\beta) = \sigma^\nu\beta$$

(cf. Lemma 5.4, p. 51). The product of any two of these automorphisms is

$$\sigma_\mu(\sigma_\nu(\beta)) = \sigma_\mu(\varepsilon^\nu\beta) = \varepsilon^{\mu+\nu}\beta = \sigma_{\mu+\nu}(\beta)$$

or briefly

$$\sigma_\mu\sigma_\nu = \sigma_{\mu+\nu}$$

where two σ's are the same if their indices are congruent mod n. Clearly the σ's constitute a cyclic group, the generator being σ_1, for we have $\sigma_\nu = \sigma_1^\nu$. The group of E/F is cyclic.

The condition that F contains a primitive n^{th} root of unity must be kept in mind. Note, for example, that $\cos 2\pi/7$ can be obtained as the root of a cubic equation with a cyclic splitting field over R. If we take the ground field $R(\omega)$ where $\omega = -1/2 + (\sqrt{3}/2)i$ is one of the primitive cube roots of unity, then $\cos 2\pi/7$ can be obtained in terms of a single root extraction in $R(\omega)$. However, we note that this is the famous *casus irreduciblis* of Cardano's formula. The equation for $\cos 2\pi/7$ has three real roots, yet it is impossible to express $\cos 2\pi/7$ in terms of real radicals.

In summary, we have proved:

THEOREM 7.1 *Given a field F containing a primitive n^{th} root of unity, an extension field E is normal and cyclic of degree n over F if and only if E is the splitting field of an irreducible equation of the form*

$$x^n - b = 0.$$

This condition is equivalent to the requirement that E be an extention of F by means of a single adjunction, $E = F(\beta)$, where β is an nth root of an element of F and no power of β less than n is in F.

LEMMA 7.2 *If F is a field containing a primitive n^{th} root of unity and $E = F(\sqrt[n]{b})$, $b \in F$, then E is the splitting field of the equation*

$$x^n - b = \prod_{i=1}^{n/r}(x^r - a_i) = 0 \quad (a_i \in F)$$

where r is the least power of $\sqrt[n]{b}$ which is contained in F.

LEMMA 7.3 *If p is prime and F is any field whatever which does not have the characteristic p, then a necessary and sufficient condition that the polynomial $x^p - b$ be irreducible is that b is not a pth power of any element in F.*

In investigating the equation $x^n - b = 0$ we have found it useful to go to a field containing the nth roots of unity. Certainly by taking a sufficiently large field we can solve any equation. Perhaps by taking a suitable extension of the ground field it may be possible to obtain great simplification in the solution of any problem. In any case, we ought to determine what gains may be derived from this method.

Let $f(x)$ be a separable polynomial over F and let $\overline{F}$ be any extension of F in which we hope for some simplification of the solution. Denote by $\overline{E}$ the splitting field of $f(x)$ over $\overline{F}$. We write $f(x)$ as a product of linear factors in $\overline{E}$,

$$f(x) = (x - \alpha_1)(x - \alpha_2)\cdots(x - \alpha_n).$$

The splitting field E of $f(x)$ over F is obtained simply by adjunction of the roots

$$E = F(\alpha_1, \alpha_2, \ldots, \alpha_n).$$

The relation between the fields is given by the scheme in the figure below. Denote the groups of E/F and $\overline{E}/\overline{F}$ by G and $\overline{G}$, respectively. What relation exists between G and $\overline{G}$?

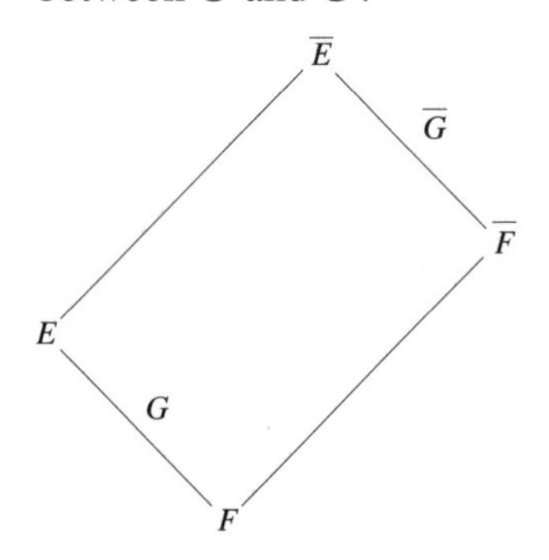

An element $\bar{\sigma}$ of $\overline{G}$ is an automorphisms of $\overline{E}$ that leaves all $\overline{F}$ fixed and consequently all F. Now each $\bar{\sigma}$ is defined by a permutation of the roots of $f(x)$. Consequently, $\bar{\sigma}$ maps the field $E = F(\alpha_1, \ldots, \alpha_n)$ onto itself. Thus each $\bar{\sigma}$ provides an automorphism of E in which F is fixed. In this way we associate an element of G with every element of $\overline{G}$. Furthermore, only one element of G is determined by any $\bar{\sigma}$ since $\bar{\sigma}$ and its image are both uniquely determined by the same permutation of the roots.

Now if two successive automorphisms $\bar{\sigma}, \bar{\tau}$ produce a certain permutation of the roots, then clearly so do their images, i.e.,

$$\bar{\sigma}\bar{\tau} = \bar{\rho} \Leftrightarrow \sigma\tau = \rho.$$

We conclude that $\overline{G}$ is isomorphic to a subgroup S of G. It is easy enough to see which subgroup it is. We shall describe S by determining the field to which it belongs.

Let Ω be the field corresponding to S. Ω is some field between E and F, and it consists of exactly these elements of E which are left fixed by S and hence consists of these elements of E left fixed by $\overline{G}$. But $\overline{G}$ leaves no other elements fixed than those of $\overline{F}$. Consequently, Ω consists exactly of those elements of E which are also in $\overline{F}$. Ω is the *intersection* or common part of the two fields. (We write $\Omega = E \cap \overline{F}$.) A schematic diagram is shown.

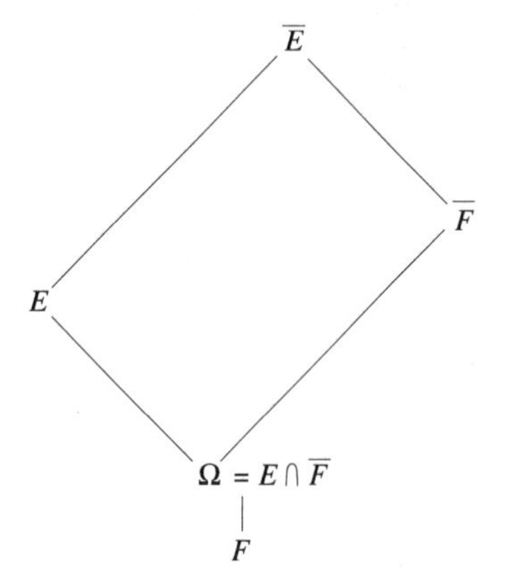

In a measure this result is disappointing. The extension $\overline{F}$ is helpful only insofar as it contains a part of $\overline{E}$. Whatever economy is achieved by the introduction of $\overline{F}$ already could have been achieved with Ω. However, we may console ourselves with the knowledge that we have eliminated any possibility of introducing mysterious solutions.

REMARK. Of course, there are other methods of obtaining roots than algebraic extension. In the field of real numbers, for example, every fifth degree equation has a root, the root being defined by some limiting process. If we adjoin one element to this field, the root of the equation $x^2 + 1 = 0$, then *all* equations are immediately solvable in the ground field. How does the introduction of limits tie in with solving equations?

We shall not digress to answer this question here but leave it as a provocative query.

Suppose the splitting field of an equation is cyclic. If the ground field does not contain n distinct n^{th} roots of unity, then these may be adjoined except in the case where the characteristic is a divisor of n. In that instance it is useless to hope for a primitive n^{th} root of unity. For if p is the characteristic we may write $n = p^{\nu}m$ where $p \nmid m$. But we have

$$x^n - 1 = x^{p^{\nu}m} - 1 = (x^m - 1)^{p^{\nu}}$$

(proposition on p. 57). Thus the nth roots of unity are the same as the mth. For a solution in terms of radicals we must therefore assume $(n, p) = 1$. In that case the equation $x^n - 1 = 0$ plainly has n distinct roots.

However, something remains to be said for the case when the characteristic is a divisor of the degree of the splitting field. If the group is cyclic, we do not expect a solution in terms of radicals, but the solution remains simple.

THEOREM 7.4 *Consider a field F of characteristic p. An extension field E is normal of degree p over F if and only if E is the splitting field of an irreducible equation of the form*

$$x^p - x - a = 0. \qquad [1]$$

This condition is equivalent to the requirement that E be an extension of F by means of a single adjunction $E = F(\alpha)$ where α is a root of the equation [1].

The root α behaves somewhat like the radical in Theorem 7.1. We shall occasionally refer to such elements as "modified radicals."

PROOF:

(1) If $x^p - x - \varepsilon$ is irreducible in F, then the splitting field E is obtained by the adjunction of a single root. It follows that $(E/F) = p$ and therefore that the group is cyclic.

The roots of the polynomial

$$f(x) = x^p - x - a$$

are distinct since $f'(x) = -1$. The splitting field must therefore be normal. Now $f(x)$ is periodic of period 1, for

$$f(x+1) = (x+1)^p - (x+1) - a = x^p + 1 - x - 1 - a = f(x).$$

Consequently, if α is a root, $f(\alpha) = 0$, the other roots are obtained simply by repeated addition of 1:

$$\alpha,\ \alpha + 1,\ \ldots,\ \alpha + p - 1.$$

Having p roots, we have all.

The splitting field of $f(x)$ is obviously $F(\alpha)$. Since $F(\alpha) = F(\alpha + \mu)$, each root $\alpha + \mu$ must satisfy an irreducible equation of the same degree as α over F. Consequently, $f(x)$ factors into polynomials of equal degree. Clearly, then, if $f(x)$ is not irreducible, it must reduce into linear factors and all the roots are already in F.

If we exclude the case where $f(x)$ has a root in F, then $f(x)$ must be irreducible. $F(\alpha)/F$ is normal of degree p and we must therefore have p automorphisms. These can only be the transformations

$$\sigma_\nu(\alpha) = \alpha + \nu \quad (\nu = 0, 1, \dots, p-1).$$

The automorphisms σ_ν clearly constitute a cyclic group.

(2) If E is normal of degree p over F (and therefore cyclic), then it is obtained by adjunction of a single element, a root of an equation of the form $x^p - x - a = 0$.

Let σ be the generator of the group. The trace

$$\theta + \sigma(\theta) + \sigma^2(\theta) + \cdots + \sigma^{p-1}(\theta)$$

cannot be zero for every $\theta \in E$ (Theorem 5.12, p. 61). Select θ so that

$$\sum_{\nu=0}^{p-1} \sigma^\nu(\theta) = b \neq 0.$$

Since $\sigma(b) = b$ it follows that $b \in F$. Setting

$$\beta = \sum_{\nu=0}^{p-1} \nu\sigma^\nu(\theta) = \sigma(\theta) + 2\sigma^2(\theta) + \cdots + (p-1)^{p-1}(\theta)$$

we obtain

$$\begin{aligned}\sigma(\beta) &= \sigma^2(\theta) + 2\sigma^3(\theta) + \cdots + (p-1)\sigma^{p-1}(\theta) \\ &= \sigma(\theta) + 2\sigma^2(\theta) + 3\sigma^3(\theta) + \cdots + (p-1)\sigma^{(p-1)}\theta \\ &\quad - [\theta + \sigma(\theta) + \sigma^2(\theta) + \sigma^3(\theta) + \cdots + \sigma^{(p-1)}(\theta)],\end{aligned}$$

that is,

$$\sigma(\beta) = \beta - b.$$

If we then set $\alpha = -\beta/b$ we obtain

$$\sigma(\alpha) = -(\beta)/b = (-\beta + b)/b = -\beta/b + 1 = \alpha + 1.$$

We have constructed an element α for which

$$\sigma^\nu(\alpha) = \alpha + \nu \quad (\nu = 0, 1, \dots, p-1).$$

The p images of α are plainly distinct and consequently α is a root of an irreducible polynomial of degree p over F (Theorem 5.16, p. 66). We conclude that $E = F(\alpha)$. It is only necessary to show that α satisfies an equation of the form $x^p - x - a = 0$.

Set $a = \alpha^p - \alpha$. We have

$$\sigma(a) = (\alpha + 1)^p - (\alpha + 1) = \alpha^p + 1 - \alpha - 1 = a.$$

Since $\sigma(a) = a$ it follows that $a \in F$. Thus α is a root of the equation

$$x^p - x - a = 0. \qquad \square$$

EXAMPLE. The polynomial $f(x) = x^5 - x - 1$ is irreducible in the rational field R. For, if $f(x)$ were reducible in R it would certainly be reducible in R_5. But, from the foregoing, we see that $x^5 - x - a$ is reducible in R_5 only if $a \equiv 0 \pmod 5$.

The general case in which the characteristic of a cyclic field is a divisor of the degree is handled in stages. Suppose E/F is normal and cyclic of degree $n = p^{\nu}m$ where p is the characteristic and $p \nmid m$. Let σ be the generator of the group. The element $\tau = \sigma p^{\nu-1}m$ generates a cyclic subgroup S of order p. The group S corresponds to a field Ω_1 which is normal and cyclic of degree $p^{\nu-1}m$ over F (cf. p. 73 ff.). By repeating this process we arrive at a chain of cyclic normal fields

$$E = \Omega_0 \supset \Omega_1 \supset \cdots \supset \Omega_{\nu} \supset F$$

where $(\Omega_i/\Omega_{i+1}) = p$ $(i = 0, \ldots, \nu - 1)$ and $(\Omega_{\nu}/F) = m$. It follows that we can obtain E from F by adjunction of mth roots of unity and one other element, some combination of radicals and modified radicals.

7.3. Steinitz' Theorem

The consideration of Theorems 7.1 and 7.4 leads us to ask what are the conditions that an algebraic extension field be obtainable by a single adjunction. (Of course, it is not always an advantage to employ a single adjunction. The field $R(\sqrt[4]{2}, i)$ is better understood in that representation than as $R(i + \sqrt[4]{2})$.) A beautiful and complete answer has been provided by Steinitz:

THEOREM *A necessary and sufficient condition that a given field of finite degree be generated by the adjunction of a single element is that there exist only finitely many fields intermediate between the given field and the ground field.*

REMARKS. The theorem is connected in some measure with the notion of separability. For the splitting field of a separable equation, the intermediate fields correspond to the subgroups of the Galois group and the number of intermediate fields must of necessity be finite. We have proved nothing for any other case. We shall, in fact, give an example where an infinite number of subfields appear.

PROOF OF NECESSITY: If a field of finite degree is generated by a single adjunction, then the number of intermediate fields is finite.

Assume $E = F(\alpha)$ and let Ω be an intermediate field, $E \supset \Omega \supset F$. It follows that $E = \Omega(\alpha)$. The element α is algebraic, E being of finite degree, since the number of independent elements $\alpha, \alpha^2, \alpha^3, \ldots$ is bounded. Thus we have a polynomial equation for α and, in particular, α must satisfy an irreducible relation $f(x) = 0$ over F. Furthermore, α must satisfy an irreducible relation $P(x) = 0$ over Ω where $P(x)$ is one of the irreducible factors of $f(x)$ over Ω. Since $E = \Omega(\alpha)$ we have $(E/\Omega) = \partial[P(x)]$ (see Lemma 5.9, p. 59).

Let Ω_0 be the field obtained by adjoining all the coefficients of $P(x)$ to F. We are sure that $F \subset \Omega_0 \subset \Omega$ since we have not adjoined any elements but those of Ω. $P(x)$ is an irreducible polynomial over Ω_0 since it is irreducible over Ω. But $(E/\Omega_0) = \partial[P(x)] = (E/\Omega)$ and $\Omega \supset \Omega_0$. We conclude that $\Omega = \Omega_0$ (see proposition on p. 60).

Any field Ω between F and $F(\alpha)$ is determined as the extension of F by the coefficients in Ω of the irreducible equation for α. Since $P(x)$ must be a divisor of $f(x)$ in E and $f(x)$ can have only a finite number of divisors, there are only finitely many possibilities for Ω. □

EXAMPLE. Consider the field $R(\sqrt[4]{2})$. The element $\sqrt[4]{2}$ satisfies the irreducible equation

$$f(x) = x^4 - 2 = 0$$

over R. In $R(\sqrt[4]{2})$, $f(x)$ has the factorization

$$f(x) = (x - \sqrt[4]{2})(x + \sqrt[4]{2})(x^2 + \sqrt{2}).$$

Using the fact that the field of the product of any two factors is the same as the field of the remaining factor, we obtain three different cases:

$$\begin{aligned} P(x) &= x - \sqrt[4]{2}, \quad & \Omega &= R(\sqrt[4]{2}), \\ P(x) &= x + \sqrt{2}, \quad & \Omega &= R(\sqrt{2}), \\ P(x) &= x^4 - 2, \quad & \Omega &= R. \end{aligned}$$

PROOF OF SUFFICIENCY: If (E/F) is finite and the number of fields intermediate between E and F is finite, then E may be obtained from F by a single adjunction $E = F(\alpha)$. We consider two possible cases:

(a) F consists of a finite number of elements.

Set $n = (E/F)$. It is easy to show (see Lemma 5.23, p. 76) that E contains q^n elements. We have proved (Corollary, p. 79) that the nonzero elements of E form a cyclic group with respect to multiplication. If α is the generator of the group, then $E = F(\alpha)$. The field consists simply of the powers of a single element.

(b) F contains infinitely many elements.

Since (E/F) is finite, E can be obtained from F by a finite number of adjunctions, trivially in fact, as the set of linear combinations of the basis elements. It is therefore sufficient to prove that an extension of F by means of two elements can always be obtained by the adjunction of a single element, i.e.,

$$F(\alpha, \beta) = F(\gamma)$$

for any α, β, and suitable γ in E.

Consider the elements

$$\gamma_c = \alpha + c\beta,$$

where $c \in F$. Since we have infinitely many c's we have an unlimited number of the γ_c at our disposal. There are only a finite number of fields $F(\gamma_c)$, however, since there are only a finite number of fields between E and F. Consequently, there must be a pair $c, d \in F$ such that

$$F(\gamma_c) = F(\gamma_d) \subset F(\alpha, \beta).$$

Now

$$\gamma_c, \gamma_d \in F(\gamma_c).$$

Hence,

$$(c - d)\beta = (\gamma_c - \gamma_d) \in F(\gamma_c).$$

Since $(c-d) \in F$ and $c-d \neq 0$ it follows that $\beta \in F(\gamma_c)$. Furthermore, α may be written as $\alpha = \gamma_c - c\beta$, whence also $\alpha \in F(\gamma_c)$. It follows that $F(\alpha, \beta) \subset F(\gamma_c)$. We conclude that

$$F(\alpha, \beta) = F(\gamma_c). \qquad \square$$

Upon re-examining the proof it is clear that we need choose a number of α's only one greater than the number of intermediate fields.

It is somewhat unsatisfactory that the two cases have to be treated differently. As yet, no proof uniting both cases has been found. As a matter of aesthetics it would be pleasant if such a proof were provided.

The connection of the theorem with separability is easily established.

LEMMA *If $E = F(\alpha_1, \alpha_2, \ldots, \alpha_n)$, where each α_i is a root of an irreducible separable polynomial $P_i(x)$, then there are only finitely many fields between E and F. By our theorem it follows that E can be obtained from F by a single adjunction.*

PROOF: Set $f(x) = P_1(x)P_2(x)\cdots P_n(x)$ and extend E to the splitting field Ω of $f(x)$, $\Omega \supset E \supset F$. Now Ω/F is normal since Ω is the splitting field of a separable polynomial over F. The fields between F and Ω correspond to the subgroups of the Galois group and therefore are finite in number. Hence there can be only finitely many fields between F and E. $\square$

Let us consider an example with infinitely many fields between the extension and the ground field. Since every polynomial is separable for fields of characteristic zero, the only exceptional cases are those where the characteristic is positive.

We shall employ a field of characteristic 2, $E = R_2(x, y)$, the field of rational functions over R_2.[2] A sample element would be any

$$\frac{x^2 + xy + x^5}{y^5 + y^3x^2};$$

all the coefficients are 1. For the ground field we take $F = R_2(x^2, y^2)$, i.e., the set of all rational functions where all the powers of the variables are even. What is the nature of the extension which gives E?

Consider any $\theta \in E$. We can express θ as a quotient of polynomials

$$\theta = \frac{\phi(x, y)}{\psi(x, y)}.$$

Since the characteristic is 2, the square of a polynomial is simply the sum of the squares of the separate terms and therefore we have

$$\theta = \frac{\phi(x, y)}{\psi(x, y)} = \frac{\phi(x, y)\psi(x, y)}{\psi^2(x, y)} = \frac{\phi(x, y)\psi(x, y)}{\psi(x^2, y^2)}.$$

The denominator $\psi(x^2, y^2)$ is an element of F. The numerator, being a polynomial, can be written in the form

$$\phi(x, y)\psi(x, y) = g_1(x^2, y^2) + g_2(x^2, y^2)x + g_3(x^2, y^2)y + g_4(x^2, y^2)xy,$$

where the $g_i(x^2, y^2)$ are polynomials.

[2] R_p is the field of the integers mod p.

We conclude that any $\theta \in E$ can be put in the form

$$\theta = a_1 + a_2 x + a_3 y + a_4 xy$$

where $a_\nu \in F$ ($\nu = 1, 2, 3, 4$). We have found a linearly independent basis, for if $\theta = 0$ we may take out the denominator and get a polynomial relation

$$f_1(x^2, y^2) + f_2(x^2, y^2)x + f_3(x^2, y^2)y + f_4(x^2, y^2)xy = 0.$$

But a polynomial can be zero only if all the coefficients are zero. It follows that the only linear expression for zero is the trivial one. Thus we have a linearly independent basis of four elements, $(E/F) = 4$.

It is impossible to derive E from F by a single adjunction. For if $E = F(\alpha)$ then α must be the root of an irreducible equation of fourth degree. But any $\alpha \in E$ already satisfies an equation of second degree since $\alpha^2 \in F$. Contradiction.

It must therefore be possible to find an infinite number of fields intermediate between E and F. Take any $\alpha \in E$. Since $\alpha^2 \in F$ it follows that $(F(\alpha)/F) \leq 2$. We may express any $\alpha \in E$ in the form

$$\alpha = a_1 + a_2 x + a_3 y + a_4 xy$$

and if not all three of a_2, a_3, a_4 are zero, then $\alpha \notin F$, $(F(\alpha)/F) = 2$. The elements of $F(\alpha)$ may be written in the form

$$A + B\alpha = C + Ba_2 x + Ba_3 y + Ba_4 xy$$

where a_2, a_3, a_4 are fixed with α. The proportion $Ba_2 : Ba_3 : Ba_4$ is constant for the entire field $F(\alpha)$. Hence to obtain an infinite number of fields we need only take an infinite number of proportions $a_2 : a_3 : a_4$. For example, consider the set of values $\alpha = x + y^{2n+1}$. We have a different field for each value of n with $a_2 = 1$, $a_3 = y^{2n}$, $a_4 = 0$.

7.4. Towers of Fields

We have seen that an equation which leads to a cyclic splitting field is solvable in terms of radicals and modified radicals. In general, if there is a chain of fields from the ground field F to an extension E,

$$F = F_0 \subset F_1 \subset F_2 \subset \cdots \subset F_n = E$$

where F_{j+1}/F is normal and cyclic for all j, then clearly F_n can be obtained from F by means of root extractions and modified radicals. If there is such an array of fields between E and F, we say E/F is a *tower*. Thus we have

THEOREM 7.5 *If E/F is a tower, then the elements of E are generated from F by means of root extractions, modified radicals, and the rational operations.*

EXAMPLES. Any cubic equation in the rational field can be solved in terms of radicals. For an irreducible cubic the splitting field is either of degree 3 or degree 6. In the first case the group is cyclic and can be solved in terms of cube roots of unity and one cubic radical. If the group is the group of order six, it contains one invariant subgroup of order 3. The field Ω corresponding to this group satisfies $(E/\Omega) = 3$ and therefore $(\Omega/F) = 2$. Consequently, the solution of a cubic reduces to the solution of a quadratic, the introduction of cube roots of unity, and the adjunction of a cube root of some element of Ω.

To show that the general equation of fourth degree can be solved in terms of radicals would require a more intimate study of the permutation group of order 24. But the same method would work. For the general quintic equation, however, we get the permutation group of order 120, and this cannot be broken down into a chain of cyclic invariant subgroups. This does not yet prove that a solution in radicals is impossible. So far we have only completed the positive side of the proof; if we can construct a chain of cyclic invariant subgroups, then an equation with rational coefficients is solvable in terms of radicals. It remains to prove the converse.

THEOREM 7.6 *If E/F is normal with a* commutative *group G, then E is a tower over F.*

PROOF: By taking the powers of any $\sigma \in G$, $\sigma \neq I$, we can easily pick out a cyclic subgroup S. The subgroup S corresponds to a field Ω, $E \supset \Omega \supset F$. Now Ω is normal over F since every subgroup of a commutative group is invariant. The group of Ω/F is simply the factor group G/S. Since the entire group is commutative, the factor group must again be commutative.

Repeating the process we can determine a field between Ω and F which has a cyclic group under Ω. In this fashion we can construct a tower of fields between F and E. □

COROLLARY *If E/F is normal with a commutative group, then E can be obtained from F by root extractions and modified radicals.*

Is it possible to effect some economy in generating E out of F? We employ certain roots of unity and other radicals and it is desirable to use radicals of the smallest possible index. This can be accomplished by breaking down each cyclic step into steps of prime order. For example, if σ is an element of period 12 we could take for the first step either the group of order 2 generated by σ^6 or the group of order 3 generated by σ^4. It is clear, then, that we need only use roots of unity and radicals for which the index is a prime divisor of the order of the group.

Let us consider the special case of $R(\varepsilon)$ where ε is a primitive n^{th} root of unity. If n is a power of a prime, $n = p^\mu$, then ε satisfies an irreducible equation of degree $p^{\mu-1}(p-1)$. What, in general, is the nature of the group of $R(\varepsilon)/R$? The only possible automorphisms have the form

$$\sigma_i(\varepsilon) = \varepsilon^i,$$

where we must have $(i, n) = 1$, for otherwise the period would be less than n.

EXERCISE 2. Prove that all the transformations

$$\sigma_i(\varepsilon) = \varepsilon^i, \quad (i, n) = 1,$$

are actually automorphisms of the field $R(\varepsilon)$. The group of $R(\varepsilon)/R$ is clearly commutative since we have

$$\sigma_i \sigma_k(\varepsilon) = \sigma_i(\varepsilon^k) = \varepsilon^{ik}.$$

Let us study the case $\mu = 1, n = p$, a prime. The degree of $R(\varepsilon)$ is $p - 1$. We can therefore construct the pth roots of unity by means of roots of unity and radicals of lower orders than p. Specifically, these orders may be restricted to the prime divisors of $p - 1$. Consequently, if $p - 1$ is a power of 2 we can generate the pth roots of unity from ± 1 and various square roots. We have proved that if p is a Fermat prime, the geometrical construction of the regular polygon of p sides is possible.

We have proved that if a field is a tower it can be generated by radicals and modified radicals. We require an approximate converse:

THEOREM 7.7 *Let $f(x)$ be an irreducible separable polynomial over the ground field F. If it is possible to give one root α of $f(x)$ by means of root extractions and rational operations, then the splitting field of $f(x)$ is a tower over F.*

PROOF: We may assume that the index of each radical is a prime since any radical can be expressed in terms of radicals of prime order; e.g.,

$$\sqrt[12]{} = \sqrt{\sqrt{\sqrt[3]{}}}.$$

If the characteristic of F is p, then we can get rid of all p^{th} roots in the expression for α. For $F(\alpha^p)$ is a subfield of $F(\alpha)$ and α satisfies the equations

$$f(x) = 0,$$
$$x^p - \alpha^p = (x - \alpha)^p = 0,$$

in $F(\alpha^p)$. Since α is a simple root of $f(x)$, the greatest common divisor of $f(x)$ and $x^p - \alpha^p$ is $x - \alpha$. But both polynomials are in $F(\alpha^p)$ and so, therefore, is their greatest common divisor. We conclude that $\alpha \in F(\alpha^p)$; i.e., α can be written as a polynomial expression in α^p. Now, if a pth root occurs in α it can be eliminated in α^p. We need only employ the relations

$$(a + b)^p = a^p + b^p \quad \text{and} \quad (ab)^p = a^p b^p$$

together with the fact that $\sqrt[m]{\sqrt[n]{}} = \sqrt[n]{\sqrt[m]{}}$. Using the polynomial relation between α and α^p we get a new expression for α. By repeating this process we can rid the expression for α of all radicals of index p. In other words, if an equation is solvable by radicals it can be solved without the use of pth roots. We assume, therefore, that p^{th} roots do not appear in the expression for α.

If $p_1, p_2, \ldots, p_s$ are the distinct indices of the radicals appearing in the expression for α, we adjoin to F the $(p_1)^{\text{th}}, (p_2)^{\text{th}}, \ldots, (p_s)^{\text{th}}$ roots of unity. The field F_1 is clearly normal and a tower over F.

Pick out an innermost radical $\sqrt[p_1]{a}$ in the formula for α. Form the field $F_2 = F_1(\sqrt[p_1]{a})$. F_2/F is normal; namely, it is the splitting field of the polynomial

$$(x^{p_1} - a)(x^{p_1} - 1)(x^{p_2} - 1) \cdots (x^{p_s} - 1).$$

Furthermore, F_2/F_1 is cyclic (Theorem 7.1). □

We then repeat the process using the next innermost radical.

The only difficulty is that we wish to have each successive field normal over F. Suppose then that we have reached a field F_i such that

(1) F_i/F is normal,

(2) F_i/F is a tower, each step being of prime degree.

The next field is to be obtained by adjoining a radical in some $\theta \in F_i$, say $\sqrt[q]{\theta}$. Denote by G_i the group of F_i/F. F_i is the splitting field of some separable polynomial $f_i(x)$. We form the new polynomial

$$f_{i+1}(x) = f_i(x) \prod (x^q - \sigma(\theta))$$

where the $\sigma(\theta)$ are the distinct images of θ through the elements of G_i. (The inclusion of the images θ is necessary to insure that the automorphisms of F_i/F are contained in the automorphisms of the new field.)

The polynomial

$$\prod_{\sigma(\theta)} (x^q - \sigma(\theta))$$

remains fixed in all the automorphisms of F_i/F and is therefore a polynomial in F. Since q is not the characteristic, this polynomial is clearly separable. It follows that $f_{1+i}(x)$ is separable and hence that the splitting field F_{i+1} of $f_{i+1}(x)$ is normal over F. Furthermore, F_{i+1} can be reached from F_i by successively adjoining q^{th} roots of the $\sigma(\theta)$, each extension being either cyclic of prime degree or involving no change at all. We conclude that F_{i+1}/F is a tower of prime steps.

In the end we reach a field Ω which is a normal tower of prime steps over F and which contains a root α of $f(x)$. Since Ω/F is normal, $f(x)$ irreducible, it follows that $f(x)$ splits in Ω (Theorem 5.16, p. 66). Hence, if one root of an irreducible polynomial is expressible in terms of radicals, every root has such an expression.

Since $f(x)$ splits in Ω, the splitting field E must be contained in Ω. We have only to prove

LEMMA 7.8 *If Ω/F is normal and a tower, E an intermediate field with E/F normal, then E/F is a tower.*

The proof may be accomplished in two ways:

(a) The group-theoretical method:

Let $\mathfrak{G}$ be the group of Ω/F, G the group of E/F. Since E/F is normal it is determined by an invariant subgroup γ of $\mathfrak{G}$. Thus G is simply the factor group

$$G = \mathfrak{G}/\gamma.$$

Now $\mathfrak{G}$ is a *solvable* group. By this we mean that there is a sequence of subgroups

$$\mathfrak{G} \supset \mathfrak{G}_1 \supset \mathfrak{G}_2 \supset \cdots \supset I \qquad [1]$$

such that $\mathfrak{G}_{i+1}$ is an invariant subgroup with respect to $\mathfrak{G}_i$ and the factor group $\mathfrak{G}_i/\mathfrak{G}_{i+1}$ is cyclic. The lemma follows from a standard theorem of group theory:

THEOREM *If a group is solvable, then every one of its factor groups is solvable.*

PROOF: Given the group $\mathfrak{G}$ in the decomposition [1], let S be any invariant subgroup and put

$$G = \mathfrak{G}/S.$$

The group G consists essentially of the elements of $\mathfrak{G}$ together with a new equivalence relation; two elements will be called equivalent if they belong to the same coset of S. Since multiplication is preserved in the mapping of $\mathfrak{G}$ onto G (cf. Exercise 11, p. 74) it is clear that the subgroups $\mathfrak{G}_1 \supset \mathfrak{G}_2 \supset \cdots \supset I$ map onto subgroups of G,

$$G \supset G_1 \supset G_2 \supset \cdots \supset I'.$$

$\mathfrak{G}_{i+1}$ is an invariant subgroup of $\mathfrak{G}_i$. It follows if $\sigma \in \mathfrak{G}_i$, $\tau \in \mathfrak{G}_{i+1}$, and s, t are the corresponding images in G_i and G_{i+1} that $\sigma\tau\sigma^{-1} \in \mathfrak{G}_{i+1}$ and hence $sts^{-1} \in G_{i+1}$. From $\mathfrak{G}_i \xrightarrow{\text{onto}} G_i$, it follows that G_{i+1} is an invariant subgroup of G_i.

We need only prove G_i/G_{i+1} cyclic. Since $\mathfrak{G}_i/\mathfrak{G}_{i+1}$ is cyclic we may write the elements as powers of some coset of $\mathfrak{G}_{i+1}$. It follows that any element of G_i/G_{i+1} can be written as a power of the corresponding coset of G_{i+1}. □

(b) The field-theoretical method:

Ω/F is normal and a tower; i.e., we have intermediate fields

$$F \subset F_1 \subset F_2 \subset \cdots \subset F_N = \Omega,$$

with F_{i+1}/F_i cyclic of prime degree. Since E is normal it is the splitting field of a separable polynomial $f(x)$ over F. Let E_i be the splitting field of $f(x)$ over F_i. The group of E_{i+1}/F_{i+1} is isomorphic to a subgroup of E_i/F_i (cf. p. 100 ff.). The group of E_1/F_1 may be G, the group of E/F. But there must be a last i for which the group E_i/F_i is G since at the end $E_N = F_N$ and the group of E_N/F_N is I. At the next step E_{i+1}/F_{i+1} must correspond to a proper subgroup S. The field between E_i and F_i which is determined by S is simply $E_i \cap F_{i+1}$. Schematically we have Figure 7.1(a). But the degree of F_{i+1}/F_i is prime and there can be no field between F_i and F_{i+1}. We cannot have $F_i = E_i \cap F_{i+1}$ for we would then have $S = G$. Therefore $E_i \cap F_{i+1} = F_{i+1}$, whence $E_i \supset F_{i+1}$. Our scheme simplifies to the arrangement in Figure 7.1(b).

Since F_{i+1}/F_i is normal and cyclic of prime order we conclude that S is an invariant subgroup of G and that the factor group G/S, the group of F_{i+1}/F_i, is cyclic of prime order.

We now ascend the tower until we reach the next simplification, i.e., until we reach a level where E_k/F_k still has the group S but E_{k+1}/F_{k+1} has a smaller group T. By the same reasoning as above, T is an invariant subgroup of S and S/T is cyclic of prime order. Continuing in this manner we must arrive at a point where no further simplification can be attained since, at the last, the group of E/N consists of the identity alone.

Thus we obtain a chain of groups

$$[1] \qquad G = G_0 \supset G_1 \supset \cdots \supset G \supset I$$

where G_{i+1} is an invariant subgroup of G_i and G_i/G_{i+1} is cyclic of prime order.

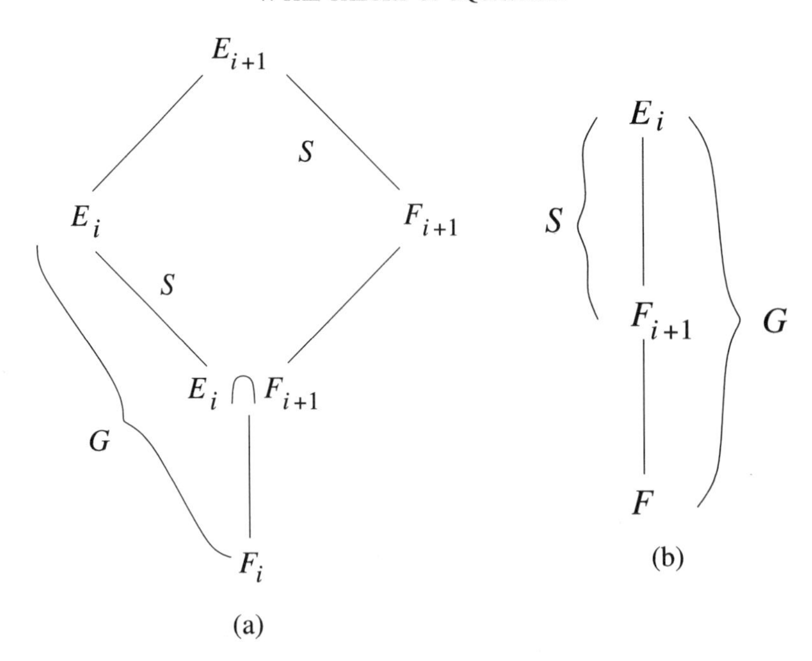

FIGURE 7.1

Consider what this means for fields between E and F. Denote the field attached to G_i by $\overline{E}_i$. G_1 is an invariant subgroup of G_0 and therefore $\overline{E}_1/F$ is normal. Furthermore, the group of $\overline{E}_1/F$ is the factor group G_0/G_1 and is cyclic of prime order. Now G_1 is the group of $E/\overline{E}_1$. The subgroup G_2 is an invariant subgroup and therefore corresponds to a field $\overline{E}_2$ normal over $\overline{E}_1$. The group of $\overline{E}_2/\overline{E}_1$ is G_2/G_1, cyclic of prime order.

In this manner we may continue by normal cyclic steps until we reach E. Clearly, then, E/F is a tower.

If the group of an equation has the form [1], then the equation is solvable by radicals. Hence any group of that form is called *solvable*.

7.5. Permutation Groups

Let F be a given field, $f(x)$ a polynomial with simple roots. If E is the splitting field of $f(x)$ we may write

$$f(x) = (x - \alpha_1)(x - \alpha_2) \cdots (x - \alpha_n) \quad (\alpha_i \in E)$$

and

$$E = F(\alpha_1, \alpha_2, \ldots, \alpha_n).$$

An automorphism σ of E is completely determined if the images of the α_i under σ are known. Now

$$\sigma(f(\alpha_i)) = f(\sigma(\alpha_i)) = 0 \Rightarrow \sigma(\alpha_i) = \alpha_k,$$

where α_k is some root of $f(x)$. Hence each σ has the effect of a permutation on the α_i. This corresponds to the original notion of Galois theory. The Galois group attached to an equation was considered not as a group of automorphisms but as a group of permutations. The permutation group of an equation is defined as the set of permutations of the roots which leaves unaltered all possible polynomial relations among the roots. Thus, a permutation σ belongs to the group if for every relation

$$P(\alpha_1, \alpha_2, \ldots, \alpha_n) = 0$$

we have also

$$P(\sigma(\alpha_1), \sigma(\alpha_2), \ldots, \sigma(\alpha_n)) = 0.$$

EXAMPLE. Consider the equation $(x^2 - 2)(x^2 - 3) = 0$. Denote the roots of the first factor by α_1, α_2, of the second by α_3, α_4. Since $\alpha_1^2 - 2 = 0$ and $\alpha_3^2 - 3 = 0$ we cannot have $\alpha_1 \to \alpha_3$ in any permutation.

The concept of the permutation which does not affect polynomial relations in the roots is much clumsier in application than that of automorphism of the splitting field. Nonetheless, there is considerable point in studying the actual permutations. By knowledge of the abstract structure of the group we could answer all field-theoretical questions. Now by considering not only the group but its effect upon particular elements we shall find it possible to answer a few questions we could not answer before.

Consider the preceding example, the equation

$$(x^2 - 2)(x^2 - 3) = 0,$$

in the rational field R. The splitting field $E = R(\sqrt{2}, \sqrt{3})$ is of degree 4 since $R(\sqrt{2}, \sqrt{3}) \supset R(\sqrt{2}) \supset R$ and $x^2 - 3$ is irreducible in $R(\sqrt{2})$ by the result of

EXERCISE 3. Show that $\sqrt{3}$ cannot be expressed in the form $a + b\sqrt{2}$ for rational a, b.

Since the degree of the splitting field is 4, all four conceivable automorphisms occur.

The field $R(\sqrt{2}, \sqrt{3})$ can also be written as $R(\sqrt{2} + \sqrt{3})$ since $R(\sqrt{2} + \sqrt{3})$ contains $1/(\sqrt{2} + \sqrt{3}) = \sqrt{3} - \sqrt{2}$. If we apply the automorphisms of the field to $\sqrt{2} + \sqrt{3}$ we get four different images and by Theorem 5.16 these are the roots of an irreducible equation of fourth degree over F, namely

$$x^4 - 10x^2 + 1 = 0.$$

Thus the same field and the same group could have been obtained from an irreducible equation.[3] From the structure of the group we cannot derive the reducibility or irreducibility of an equation! It is just the additional knowledge of the specific permutations of the roots which enables us to treat these questions.

Denote the roots of the equation $f(x) = 0$ by α_i. An automorphism of the splitting field effects a certain permutation of the roots

$$\sigma(\alpha_i) = \alpha_{\nu_i}.$$

[3]This is generally true by Steinitz' theorem.

We shall denote the roots by their subscripts alone and σ will then be defined by the notation

$$\sigma = \begin{pmatrix} 1, 2, \ldots, n \\ \nu_1, \nu_2, \ldots, \nu_n \end{pmatrix}.$$

The *product* of two permutations is defined as the result of applying them in succession. In the example of $(x^2 - 2)(x^2 - 3)$ we have

$$\begin{pmatrix} 1 & 2 & 3 & 4 \\ 2 & 1 & 4 & 3 \end{pmatrix}\begin{pmatrix} 1 & 2 & 3 & 4 \\ 2 & 3 & 4 & 1 \end{pmatrix} = \begin{pmatrix} 1 & 2 & 3 & 4 \\ 1 & 4 & 3 & 2 \end{pmatrix}.$$

In this instance the permutation group does not carry every digit over into every other digit since, e.g., α_1 cannot go into α_3. If, on the contrary, a group of permutations carries every digit over into every other digit, we say the group is *transitive.*

LEMMA 7.9 *A necessary and sufficient condition that a group of permutations be transitive is that the digit* 1 *can be carried into any other; i.e., as* σ *runs through the group,* $\sigma(1)$ *runs through the digits* $1, 2, \ldots, n$.

PROOF: The condition is obviously necessary. In order to prove it is sufficient, we show it is possible to carry any digit j over into any other digit k. Now j and k both appear as images of 1. Thus there are permutations σ, τ with

$$\sigma(1) = j, \quad \tau(1) = k.$$

In the inverse to σ we have $\sigma^{-1}(j) = 1$, whence $\tau(\sigma^{-1}(j)) = k$. □

Suppose, for a given group G, that 1 is carried over into the digits $1, 2, \ldots, r$ and no others. (This is no restriction since we may label our elements so that the images of 1 are the first r indices.) By the proof of Lemma 7.9 it follows that any one of these digits can be mapped onto any other. Furthermore, no permutation of the group will move any one of the digits $1, 2, \ldots, r$ into a digit $k > r$. For assume $j \leq r$ with $\sigma(j) = k$. Since j is an image of 1, $j = \tau(1)$, we have $\sigma\tau(1) = k$.

The group permutes r of the digits in a transitive way. The set of digits $1, 2, \ldots, r$ is called a *domain of transitivity.* We may divide all the digits into domains of transitivity, a domain of transitivity consisting of all the integers that can be carried into each other by the permutation of the group.

LEMMA 7.10 *There is a one-to-one correspondence between the irreducible factors of a separable polynomial* $f(x)$ *and the domains of transitivity of its Galois group.*

PROOF: Let

$$P(x) = (x - \alpha_1)(x - \alpha_2) \cdots (x - \alpha_r)$$

be an irreducible factor of $f(x)$ over F. From $P(\alpha_1) = 0$ we have $P(\sigma(\alpha_1)) = 0$ where σ is any element of the group of $f(x)$. Consequently, $\sigma(\alpha_1)$ is one of $\alpha_1, \alpha_2, \ldots, \alpha_r$. Furthermore, α_1 has at least r distinct images for otherwise it would satisfy an equation of lower degree (Theorem 5.16, p. 66). It follows that each α_i, $i \leq r$, is an image of α_1 in some automorphism, and α_1 has no other images. In other words, the irreducible factor $P(x)$ determines the transitivity domain $1, 2, \ldots, r$. □

COROLLARY *The group of an irreducible separable polynomial is transitive.*

EXAMPLE. The field $R(\sqrt{2}, \sqrt{3}) = R(\sqrt{2} + \sqrt{3})$ may be considered as the splitting field of either $x^4 - 5x^2 + 6$ or $x^4 - 10x^2 + 1$. The abstract group of both polynomials is the same and corresponds to the automorphisms $\sqrt{2} \to \pm\sqrt{2}$, $\sqrt{3} \to \pm\sqrt{3}$. We denote these automorphisms by I, σ, τ, ρ according to the schedule

	I	σ	τ	ρ
$\sqrt{2} \to$	$\sqrt{2}$	$-\sqrt{2}$	$\sqrt{2}$	$-\sqrt{2}$.
$\sqrt{3} \to$	$\sqrt{3}$	$\sqrt{3}$	$-\sqrt{3}$	$-\sqrt{3}$

The effect of the group on the roots of both equations is given in the tables

$x^4 - 5x^2 + 6$

$$\alpha_1 = \sqrt{2}, \quad \alpha_2 = -\sqrt{2}$$
$$\alpha_3 = \sqrt{3}, \quad \alpha_4 = -\sqrt{3}$$

	1 2 3 4
I	1 2 3 4
σ	2 1 3 4
τ	1 2 4 3
ρ	2 1 4 3

$x^4 - 10x^2 + 1$

$$\alpha_1 = \sqrt{2} + \sqrt{3}, \qquad \alpha_2 = \sqrt{2} - \sqrt{3}$$
$$\alpha_3 = -\sqrt{2} + \sqrt{3}, \quad \alpha_4 = -\sqrt{2} - \sqrt{3}$$

	1 2 3 4
I	1 2 3 4
σ	2 1 4 3
τ	3 4 1 2
ρ	4 3 2 1

In the first case we have two domains of transitivity, each containing two elements. In the second case we obtain a transitive group of order four. The situation is entirely different yet the structure of the abstract group is the same in both cases.

We are particularly interested in groups which are not solvable. Apparently most groups are solvable, the smallest nonsolvable group being of order 60. The nonsolvable group of order 60 is a simple group; a *simple group* is a group which has no invariant subgroups other than itself and the identity. Clearly, every cyclic group of prime order is simple. Apart from these, the simple groups seem to be very rare. The next nontrivial simple group is of order 168 and is given by the symmetries of the abstract geometry of seven points.

This is a projective geometry defined by the postulates:

(a) there exists at least one line;
(b) every line contains exactly three points;
(c) there is at least one point not on a given line;
(d) two points lie on exactly one line;
(e) two lines intersect in exactly one point.

This geometry is represented in the table below and in the accompanying figure:

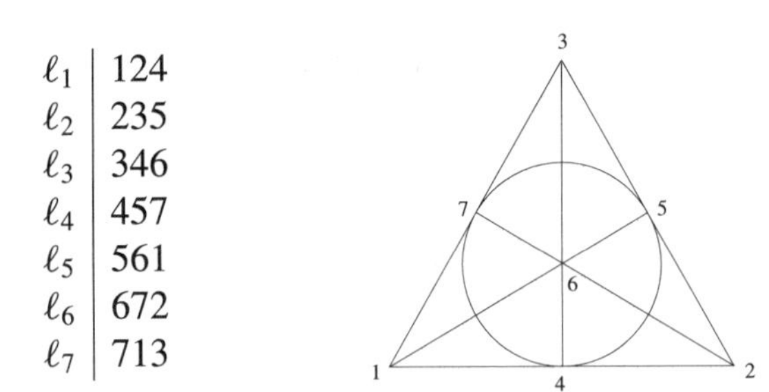

ℓ_1	124
ℓ_2	235
ℓ_3	346
ℓ_4	457
ℓ_5	561
ℓ_6	672
ℓ_7	713

The simple group of 168 elements consists of all permutations of the digits $1, 2, \ldots, 7$ for which the collineation relations of this geometry remain unchanged. Thus, for example, the cyclic permutations belong to this group.

EXERCISE 4. Determine all the permutations of the simple group of order 168.

The simple group of order 60 also has a geometrical interpretation. It is the group of rotations of the icosahedron.

There has not yet been any solution to the problem of determining all simple groups.[4] The first few have the orders 60, 168, 360, 504, 660, 1092, there being one simple group for each order. However, there may be more than one simple group of a given order; there are two of order 20160. The most common simple groups appear to be those of order $p(p^2-1)/2$ where p is any prime greater than 3. In particular, the group of the matrices $\left(\begin{smallmatrix} a & b \\ c & d \end{smallmatrix}\right)$ with elements in R_p and determinant equal to 1 has a factor group of order $p(p^2-1)/2$ with respect to the invariant subgroup

$$\begin{pmatrix} 1 & 0 \\ 0 & 1 \end{pmatrix}; \quad \begin{pmatrix} -1 & 0 \\ 0 & -1 \end{pmatrix}$$

and this is simple.

It is probable that every group appears as the group of some equation over the rational field but most equations have certain special groups. The general equation of nth degree leads to the so-called *symmetric group* of order $n!$. This is the group of all permutations on n objects. In general, the symmetric group is nonsolvable as we shall prove.

A cyclic permutation will now be described by writing the digits in the cyclic order, e.g.,

$$\begin{pmatrix} 1 & 2 & 3 & 4 & 5 & 6 & 7 \\ 2 & 3 & 4 & 5 & 6 & 7 & 1 \end{pmatrix} = \begin{pmatrix} 1 & 2 & 3 & 4 & 5 & 6 & 7 \end{pmatrix}.$$

Such a permutation will be referred to briefly as a *cycle*.

PROPOSITION *Every permutation can be written as a product of disjoint cycles.*

EXAMPLE.

$$\sigma = \begin{pmatrix} 1 & 2 & 3 & 4 & 5 & 6 & 7 \\ 2 & 3 & 1 & 4 & 6 & 7 & 5 \end{pmatrix} = (123)(4)(567).$$

[4] *Editor's note*: This is now known.

We agree to omit mention of any digit that remains fixed, e.g.,

$$\sigma = (123)(567).$$

If we employ the special convention of writing the identity as I every permutation can be described in this form.

EXAMPLES. The permutation group on 3 objects consists of

the identity:	I
2-cycles:	(12), (13), (23)
3-cycles:	(123), (132).

The permutation group on 4 elements consists of

the identity:	I
2-cycles:	(12), (13), (14), (23), (24), (34)
products of 2-cycles:	(12)(34), (13)(24), (14)(23)
3-cycles:	(123), (124), (132), (134), (142), (143), (234), (243)
4-cycles:	(1234), (1243), (1324), (1342), (1423), (1432)

with 24 permutations in all.

The period of a cycle is clearly equal to its length. In the powers of any permutation the disjoint cycles may behave independently. It follows that

LEMMA 7.11 *The period of a permutation is the least common multiple of the lengths of its constituent cycles.*

The following rule of computation will be found useful:
If τ is given in cycle decomposition, the transform $\sigma\tau\sigma^{-1}$ is obtained simply by performing the permutation σ on the separate cycles of τ. Thus if

$$\tau = (314)(25)(67)$$
$$\sigma = (123)(567)$$

then

$$\sigma\tau\sigma^{-1} = (124)(36)(75).$$

There is another representation which is very useful. Every permutation can be written as a product of *transpositions* (2-cycles) but here the terms are not necessarily disjoint. Thus, in particular, we have for a cycle

$$(0, 1, 2, \dots, n) = (0, n)(0, n - 1) \cdots (0, 1).$$

Note. An $(n + 1)$–cycle can be written as the product of n transpositions. The representation as a product of transpositions for any permutation can be obtained by writing each of its disjoint cycles as above. This is not the only conceivable way of writing a permutation as a product of transpositions but all such representations have something in common.

LEMMA 7.12 *In all the representations of a permutation as a product of transpositions the number of transpositions has the same parity; i.e., the number of transpositions is either always even or always odd.*

PROOF: Construct the polynomial in the n variables $x_1, x_2, \ldots, x_n$ consisting of the product of their differences,

$$P = \prod_{1 \leq i < j \leq n} (x_i - x_j),$$

or, in expanded form,

$$\begin{aligned} P = (x_1 - x_2)(x_1 - x_3) \cdots (x_1 - x_n) \\ (x_2 - x_3) \cdots (x_2 - x_n) \\ \vdots \\ (x_{n-1} - x_n). \end{aligned}$$

If we permute the indices of the x_i, then each factor goes into some other factor or its negative. Thus a permutation has either the effect of changing the sign of P or leaving it the same. □

EXERCISE 5. Show that the effect of a transposition is to change the sign of P.

If a permutation leaves the sign of P unchanged, it is said to be *even*; otherwise it is called *odd*. The product of two even permutations or two odd permutations is even. The product of an even and an odd is odd. In view of Exercise 5 it is clear that an even number of transpositions is required to represent an even permutation, an odd number of transpositions to represent an odd.

COROLLARY *The inverse of a permutation is even or odd according to whether the original permutation is even or odd.*

The even permutations, being closed under multiplication and taking inverses, form a subgroup of the full symmetric group. The subgroup of the even permutations is clearly invariant:

$$\sigma S \sigma^{-1} = S.$$

The order of the subgroup is easily determined:

There are as many even permutations as odd, for the products of the distinct even permutations with any fixed odd permutation are distinct odd permutations. Hence the number of odd permutations is at least as great as the number of even permutations. But if we multiply the distinct odd permutations by any single odd permutation we get distinct even products. Hence the number of even permutations is not less than the number of odd. We conclude that the even permutations are half the symmetric group. They constitute an invariant subgroup of order $\frac{1}{2}n!$. This is the so-called *alternating* group on n elements.

The symmetric group contains an invariant subgroup of order $\frac{1}{2}n!$ and index 2. The corresponding factor group must then be cyclic. But this, in general, is as far as the decomposition of the symmetric group can be carried. Except in the cases $n \leq 4$, the alternating group on n elements is nonsolvable.

THEOREM 7.13 *The alternating group on n elements is a simple group.*[5]

The proof will require the development of some preliminary results.

Denote the symmetric group on n elements by S_n, the alternating group by A_n. If we consider the cycle decomposition of a permutation, we can tell at once whether it is even or odd, a cycle in an odd number of elements being even, otherwise odd (see note on p. 117). Thus A_3 consists of the identity and the 3-cycles, 3 elements in all; A_4 consists of the identity, the products of 2-cycles, and the 3-cycles, 12 elements in all.

PROPOSITION 7.14 *If $G \neq I$ is an invariant subgroup of A_n or S_n, $n \neq 4$, then G contains either a 2-cycle or a 3-cycle.*

PROOF: Select a $\tau \in G$ whose period is a prime p, $\tau^p = I$. For this purpose we need only pick out any $\sigma \neq I$. If the period of σ is not a prime, then it has a prime divisor and this is the period of some power of σ.

By Lemma 7.11 (p. 117), the cycle decomposition of τ can only contain p-cycles. Hence, we have only the following possibilities for τ:

(1) $\tau = (12)$, a 2-cycle, in which case we are finished.
(2) $\tau = (12)(34)\cdots$, a product of 2-cycles.
(3) $\tau = (123)$ a 3-cycle, in which case we are done.
(4) $\tau = (123)(456)\cdots$, a product of 3-cycles.
(5) $\tau = (12345 \text{ or more})\cdots$, a product of p-cycles, $p > 3$.

We shall have to dispose of the cases (2), (4), and (5). G is assumed to be an invariant subgroup of A_n or S_n. Hence if $\tau \in G$, $\sigma \in A_n$ we have

$$\sigma\tau\sigma^{-1} \quad \text{and} \quad \sigma\tau^{-1}\tau^{-1} \in G.$$

Case 4. $\tau = (123)(456)\cdots$.
Taking $\sigma = (1234)$ we have by the rule on page 117

$$\begin{aligned} \sigma\tau\sigma^{-1} &= (134)(256) \text{ (all other terms remain the same)}, \\ \sigma\tau\sigma^{-1}\tau^{-1} &= (14235)(6)I \\ &= (14235). \end{aligned}$$

Thus we have reduced case 4 to

Case 5. $\tau = (12345\cdots p)\cdots$.
Taking $\sigma = (234)$ we obtain $\sigma\tau\sigma^{-1} = (13425\cdots p)\cdots$, the dots indicating elements which remain the same,

$$\begin{aligned} \sigma\tau\sigma^{-1}\tau^{-1} &= (1)(352)(4)(6)\cdots(p)I \\ &= (352). \end{aligned}$$

Hence if G contains an element of the type of 4 or 5 it contains a 3-cycle. It remains to complete the discussion of

Case 2. $\tau = (12)(34)\cdots$.

[5] *Editor's note*: This is true only for $n > 4$.

Taking $\sigma = (123)$ we have

$$\begin{aligned}\sigma\tau\sigma^{-1} &= (23)(14) \text{ (all other cycles remain the same)},\\ \sigma\tau\sigma^{-1}\tau^{-1} &= (13)(24)\,\mathrm{I}\\ &= (13)(24).\end{aligned}$$

There are two possibilities:

$n > 4$. If G contains a product of 2-cycles $\tau = (12)(34)$, then, taking $\sigma = (125)$, we have

$$\begin{aligned}\sigma\tau\sigma^{-1} &= (25)(34),\\ \sigma\tau\sigma^{-1}\tau^{-1} &= (152)(3)(4) = (152),\end{aligned}$$

and G contains a 3-cycle.

$n = 4$. Set $\tau = (12)(34), \sigma = (123)$. We obtain

$$\sigma\tau\sigma^{-1} = (23)(14), \quad \sigma\tau\sigma^{-1}\tau^{-1} = (13)(24).$$

Hence, G contains three elements of the same form. These three elements together with the identity form an invariant subgroup—the familiar four-group $\mathcal{F}$. □

COROLLARY *The symmetric group on four elements is a solvable group.*

PROOF: S_4 contains an invariant subgroup A_4 which contains $\mathcal{F}$ as an invariant subgroup. Hence S_4 possesses a decomposition into invariant subgroups according to the following scheme:

$$I \underset{2}{\lhd} S_2 \underset{2}{\lhd} \mathcal{F} \underset{3}{\lhd} A_4 \underset{2}{\lhd} S_4.$$

We have succeeded in proving that *every equation of the fourth degree possesses a solution in terms of radicals.* □

PROPOSITION 7.15 *If $n \neq 4$ the only invariant subgroups of the symmetric group are A_n and S_n.*

PROOF: The cases $n = 2, 3$ are trivial. Assume $n > 4$. Now, by Proposition 7.14, if G is an invariant subgroup it must contain either a 2-cycle or a 3-cycle.

If G contains a 2-cycle, say $\sigma = (12)$, then it contains all 2-cycles. We have, e.g.,

$$(234)(12)(234)^{-1} = (13) \in G.$$

It follows that G is the whole symmetric group since every element can be represented as a product of transpositions (2-cycles).

Similarly, we can show if G contains any 3-cycle it contains all. For suppose G contains $\sigma = (123)$. It is sufficient to show that any digit can be transformed into any other, $(123) \to (124)$ say. We have $(123)^2 = (132) = \sigma^2 \in G$. Taking $\tau = (12)(34)$ we obtain

$$\tau\sigma^2\tau^{-1} = (241).$$

Thus G contains every 3-cycle. It follows that G contains A_n by

PROPOSITION 7.16 *Every even permutation can be expressed as a product of 3-cycles.*

PROOF: Represent the permutation in the disjoint cycle decomposition. If each cycle is decomposed according to the method of page 117, we obtain an even number of transpositions. Pair these off, the first with the second, the third with the fourth, etc. Two possible cases occur.

(a) The two transpositions in a pair may have exactly one digit in common. In that case we may write the product as a 3-cycle,

$$(12)(13) = (132).$$

(b) The two transpositions of a pair have no common digits. In that case we have

$$(12)(34) = (12)(13)(13)(34) = (132)(134). \qquad \square$$

The statement is proved. □

As a consequence of the foregoing results we have

THEOREM 7.17 *The symmetric group on n elements, $n > 4$, is not solvable.*

PROOF: It contains only one invariant subgroup, the alternating group, and this is a simple group of nonprime order. □

From this result we shall prove that there is no formula in terms of radicals which can be used to solve the general equation of nth degree; i.e., there is no formula which works for every possible choice of the coefficients. We will have yet to preclude the possibility that each equation can be solved by a special method.

7.6. Abel's Theorem

Let K be any field and form the transcendental extension

$$F = K(a_1, a_2, \ldots, a_n)$$

by means of n free variables $a_1, a_2, \ldots, a_n$. The equation

$$f(x) = x^n + a_1 x^{n-1} + \cdots + a_n = 0$$

over F possesses no solution in radicals for $n > 4$.

PROOF: Let E be the splitting field of $f(x)$. The proof will consist in showing that the group of E/F is the symmetric group on n elements.

In E we may write

$$f(x) = (x - x_1)(x - x_2) \cdots (x - x_n).$$

E is obtained from F by the adjunction of the roots of $f(x)$;

$$E = F(x_1, x_2, \ldots, x_n).$$

Since the a_i are rational expressions in the x_i,

$$[1] \qquad \begin{cases} a_1 = -(x_1 + x_2 + \cdots + x_n) \\ a_2 = +(x_1 x_2 + x_1 x_3 + \cdots + x_{n-1} x_n) \\ \vdots \\ a_n = (-1)^n (x_1 x_2 \cdots x_n) \end{cases}$$

we have

$$E = K(x_1, x_2, \ldots, x_n).$$

The x_i are not interdependent, as we shall prove. Hence the solution of the general equation of nth degree will be given by the field of rational functions of just n free variables!

Let us investigate this field. Set

$$\overline{E} = K(y_1, y_2, \ldots, y_n)$$

where the y_i are independent variables. We seek a field between $\overline{E}$ and K isomorphic to F. Trivially, all $n!$ permutations of the variables y_i are automorphisms of $\overline{E}$. Denote the fixed field of this group by $\overline{F}$. The field $\overline{F}$ certainly contains the elements

$$\text{[2]} \qquad \begin{cases} b_1 = -(y_1 + y_2 + \cdots + y_n) \\ b_2 = (y_1y_2 + y_1y_3 + \cdots + y_{n-1}y_n) \\ \quad\vdots \\ b_n = (-1)^n(y_1y_2\cdots y_n). \end{cases}$$

The field of these elements

$$\Phi = K(b_1, b_2, \ldots, b_n)$$

is a subfield of $\overline{F}$. Now $\overline{E}$ is clearly the splitting field of the polynomial

$$\bar{f}(x) = x^n + b_1x^{n-1} + \cdots + b_n = (x - y_1)(x - y_2)\cdots(x - y_n)$$

over Φ. We have $(\overline{E}/\overline{F}) = n!$ and $\overline{E} \supset \overline{F} \supset \Phi$. To prove $\Phi = \overline{F}$ we need only show that $(\overline{E}/\Phi) \leq n!$.

We have

$$\overline{E} = \Phi(y_1, y_2, \ldots, y_n).$$

If we put

$$\Phi_i = \Phi(y_1, y_2, \ldots, y_i)$$

we have a chain of fields

$$\Phi \subset \Phi_1 \subset \Phi_2 \subset \cdots \subset \Phi_n = E.$$

Now $(\Phi_1/\Phi) \leq n$ since y_1 is a root of the nth degree polynomial $\bar{f}(x)$. Similarly, since y_2 is a root of $\bar{f}(x)/(x - y_1)$ we have $(\Phi_2/\Phi_1) \leq n - 1$. In the same way

$$(\Phi_{j+1}/\Phi_j) \leq n - j$$

whence we obtain

$$(\overline{E}/\Phi) \leq n!.$$

Hence $\Phi = \overline{F}$ and $(\overline{E}/\Phi) = n!$. We have shown that

$$\overline{F} = K(b_1, b_2, \ldots, b_n).$$

In other words, every rational function in $y_1, \ldots, y_n$ which remains invariant under all permutations of the y_i is a rational function of $b_1, \ldots, b_n$, the so-called *symmetric functions* in the y_i.

The b_i must then be free variables; i.e., there can be no algebraic relation among them. Otherwise we would have a nontrivial polynomial relation over K

$$P(b_1, b_2, \ldots, b_n) = 0.$$

Substituting for the b's by means of [2] we obtain a polynomial relation

$$P(-(y_1 + y_2 + \cdots), (y_1 y_2 + \cdots), (-1)^n (y_1 y_2 \cdots y_n)) = 0.$$

Since the y_i are free variables, this relation is an identity which holds no matter what we take for the y's. Replace y_i by x_i. Then by [1] we have

$$P(x_1, a_2, \ldots, a_n) = 0.$$

But the a_i are free variables and consequently P is identically zero. Contradiction.

Since the b_i are free variables the field $\overline{F} = K(b_1, b_2, \ldots, b_n)$ is isomorphic to $F = K(a_1, a_2, \ldots, a_n)$ under the mapping $a_i \leftrightarrow b_i$. But $\bar{f}(x)$ is the image of $f(x)$ under this mapping. Since $\overline{E}$ and E are the splitting fields they must be isomorphic by an extension of the mapping of $\overline{F}$ on F (Theorem 5.3, p. 50). It follows that the group of E/F is the symmetric group on n elements. We conclude by Theorem 7.17 that the general equation of nth degree possesses no solution in radicals. □

So we must abandon any hope that all equations of nth degree can be solved in radicals by some general formula. We must even discard the possibility that each particular case can be solved by special methods.

7.7. Polynomials of Prime Degree

Let $f(x)$ be an irreducible separable polynomial over F with $\partial[f(x)] = p$, a prime. If $f(x)$ is solvable by radicals, what may we conclude about the group G? G is essentially a transitive group on p digits (Lemma 7.10). Furthermore, G is a solvable group; i.e., there is a chain of invariant subgroups

$$G = G_n \supset G_{n-1} \supset \cdots \supset G_1 = I$$

such that the orders of the factor groups G_{i+1}/G_i are prime. We shall need

LEMMA 7.18 *An invariant subgroup $H \neq I$ of a transitive group G on p elements, p prime, is again transitive.*

PROOF: Let $0, 1, 2, \ldots, k-1$ be a domain of transitivity of H. Thus, each of these digits goes over into every other in the permutations of H and no digit greater than $k-1$ appears as an image of these. Now let j be any digit whatever. Since G is a transitive group there is a $\sigma \in G$ with $\sigma(0) = j$. From $\sigma H \sigma^{-1} = H$ we conclude that $H(j) = \sigma H \sigma^{-1}(j) = \sigma H(0)$. But, since 0 goes into all digits from 0 to $k-1$ in H

$$\sigma H(0) = \sigma(0), \sigma(1), \ldots, \sigma(k-1).$$

Hence j is contained in a domain of transitivity of k digits. In other words, all domains of transitivity have the same size. It follows that $k|p$. But p is prime, $k \neq 1$, and therefore $k = p$. H must therefore be transitive. □

From this lemma we conclude that all the groups G_i, $i > 1$, in the decomposition of G must be transitive.

LEMMA 7.19 *Every permutation in G may be written as a linear transformation in the digits* $0, 1, 2, \ldots, p-1$, *i.e., a transformation*

$$z \to az + b \,(\text{mod } p)$$

whence $a \not\equiv 0$ (mod p).

PROOF: The group G_2 is both cyclic of prime order and transitive. Denote the generator of G_2 by σ. G_2 has order p and σ is a p-cycle. For if σ had a cycle of shorter length in the representation as a product of disjoint cycles, then the digits of that cycle would form a domain of transitivity in G. Consequently, by labeling the digits properly, we may set

$$\sigma = (0, 1, 2, \ldots, p-1).$$

The elements of G_2 are the transformations

$$\sigma^{\nu} = (z \to z + \nu).$$

These transformations, we shall see, are the only linear transformations which leave no digit fixed. Hence, if the lemma is true, the only p-cycles in G are elements of G_2. What are the fixed digits of the transformation $z \to az + b$? These are solutions of the equation

$$z \equiv az + b \text{ (mod } p)$$

or, equivalently, of

$$(a-1)z \equiv -b \text{ (mod } p).$$

There are two cases:

Case 1. $a \not\equiv 1$.
In this case the equation always has a solution and it is unique.

Case 2. $a \equiv 1$.
The transformation has the form

$$z \to z + b.$$

This is the permutation σ^b. It has either all digits fixed or no digits fixed. Thus in a group of linear transformations the p-cycles can have only this form.

Now, if the lemma is true for G_i it is true for G_{i+1}. First of all, the only p-cycles in G_i are elements of G_2. Hence if $\tau \in G_{i+1}$ then $\tau\sigma\tau^{-1}$ is a p-cycle (see rule on p. 117) in G_i, $\tau\sigma\tau^{-1} = \sigma^b$. We then have

$$\tau\sigma = \sigma^b\tau.$$

Let us consider the effect of τ on any digit k:

$$\tau\sigma(k) = \sigma^b\tau(k)$$

whence

$$\begin{aligned}
\tau(k+1) &= \tau(k) + b, \\
\tau(k+2) &= \tau(k+1) + 1 = \tau(k) + 2b, \\
&\vdots \\
\tau(k+z) &= \tau(k) + zb.
\end{aligned}$$

Setting $k = 0$ we have

$$\tau(z) = \tau(0) + zb$$

or, in the original notation,

$$\tau(z) = bz + c$$

where $c = \tau(0)$.

Thus we have proved the lemma for G_{i+1} if it is true for G_i. Since the lemma is true for G_2 it is true for all G. We have proved for a solvable equation of prime degree that the only permissible permutations of the roots have the form

$$z \to az + b. \qquad \square$$

THEOREM 7.20 *The splitting field of a solvable equation of prime degree is generated by any pair of roots.*

PROOF: Let $\alpha_0, \alpha_1, \ldots, \alpha_{p-1}$ be the roots of $f(x)$ over F and form the field

$$E = F(\alpha_0, \alpha_1, \ldots, \alpha_{p-1}).$$

Choose any pair of α_i, α_k and form the field $\Omega = F(\alpha_i, \alpha_k)$. What is the group of E/Ω? This is the subgroup of permutations which leave α_i and α_k fixed. But, by the proof of Lemma 7.19 the only permutation which leaves two elements fixed is the identity. Hence the group of E/Ω is the identity and $E = \Omega$. $\square$

This theorem has an interesting consequence:

COROLLARY 7.21 *A solvable irreducible equation of prime degree which has two real roots has all roots real.*

COROLLARY 7.22 *If an irreducible equation is of prime degree greater than three and possesses precisely three real roots, it cannot be solvable.*

EXERCISE 6. Find an equation of fifth degree with integer coefficients and with precisely three real roots.

There are no proven results concerning the frequency with which solvable equations occur, but experience indicates that most equations have the full symmetric group. Clearly any group may occur as the Galois group of an equation provided we do not preassign the ground field. On the other hand, if the ground field is the field of rational numbers, is it always possible to determine a normal extension which has that group? The answer is probably yes, but nobody has succeeded in finding a proof. Naturally, we cannot expect such a result for any ground field. For example, every polynomial in the field of complex numbers possesses a splitting in the ground field. In the field of real numbers any equation can be solved by an extension of degree 2. Hence, there is no possible algebraic answer to the problem, but it remains an interesting question in number theory.

Titles in This Series

16 **S. R. S. Varadhan,** Stochastic processes, 2007

15 **Emil Artin,** Algebra with Galois theory, 2007

14 **Peter D. Lax,** Hyperbolic partial differential equations, 2006

13 **Oliver Bühler,** A brief introduction to classical, statistical, and quantum mechanics, 2006

12 **Jürgen Moser and Eduard J. Zehnder,** Notes on dynamical systems, 2005

11 **V. S. Varadarajan,** Supersymmetry for mathematicians: An introduction, 2004

10 **Thierry Cazenave,** Semilinear Schrödinger equations, 2003

9 **Andrew Majda,** Introduction to PDEs and waves for the atmosphere and ocean, 2003

8 **Fedor Bogomolov and Tihomir Petrov,** Algebraic curves and one-dimensional fields, 2003

7 **S. R. S. Varadhan,** Probability theory, 2001

6 **Louis Nirenberg,** Topics in nonlinear functional analysis, 2001

5 **Emmanuel Hebey,** Nonlinear analysis on manifolds: Sobolev spaces and inequalities, 2000

3 **Percy Deift,** Orthogonal polynomials and random matrices: A Riemann-Hilbert approach, 2000

2 **Jalal Shatah and Michael Struwe,** Geometric wave equations, 2000

1 **Qing Han and Fanghua Lin,** Elliptic partial differential equations, 2000